Yoshiyuki Usami

Associate professor, Ph Dr.

Kanagawa University

Rokkakubashi 3-27-1, Yokohama 221-8686, Japan

e-mail usami@kanagawa-u.ac.jp

Biomechanics for bipedal dinosaurs
How fast Tyrannosaurus could run

Preface

Chapter 1   Structure of Skeletal Muscle

Chapter 2   Muscle Contraction Property

    2.1 Isometric v.s. isotonic and twitch v.s. tetanus
    2.2 P_0 [N/m^2] "Isometric tetanus" = "Maximum isometric tension"
       = "Isometric tetanic tension"
    2.3 Hill's equation: A basic model expressing muscle contraction:
    2.4 Mathematical properties of Hill's equation
    2.5 Improved version of Hill's relation.

Chapter 3   The maximum tetanic tension P_0.

    3.1 The maximum muscle stress σ, and the maximum tetanic tension P_0
    3.2 The isometric tetanic tension P_0
    3.3 Running ability of T.rex and P_0

Chapter 4   Maximum muscle stress and specific tension σ

    4.1 Maximum muscle stress σ
    4.2 Measuring the Maximum Muscle Force σ N/cm^2
    4.3 Using MRI scanning for determining specific tension σ N/cm^2
    4.4 Why σ is so scattered
    Appendix to Chap.4

Chapter 5   Torque v.s. Muscle volume

Chapter 6   Mechanical Power Output of Muscle

    6.1 The maximum isotonic power
    6.2 Work Loop Technique
    6.3 Experiment using work loop technique
    6.4 Can power obtained by muscle contraction studies explain animal movement ?

Chapter 7   Muscle – Tendon Complex

    7.1 What role tendon plays while muscle is shortening ?
    7.2 What happen to muscle and tendon during walking?

Chapter 8   Evaluation of running ability for *T.rex*

    8.1 Estimation of running speed of *T.rex* in the literature

8.2 Alexander's qualitative speed evaluation method
8.3 How we estimate hip height h from foot print?
8-4 Computer simulation studies

Chapter 9 Casting a doubt for fast running ability of *T.rex*

9.1 Was *T.r*ex not a fast runner?
9.2 Basic biomechanics for the static theory
9.3 Uncertainty of the parameters L: muscle fibre length
9.4 Uncertainty of the parameters θ: pennation angle
9.5 Uncertainty of the parameters r: moment arm

Chapter 10   Dynamical calculation of the locomotion of *T.rex*

10.1 Dynamical calculation of the locomotion of *T.rex*
10.2 Evolutionary computation method
10.3 Simulation Results - Running Motion Generated by Evolutionary Algorithm-
10.4 Obtained parameters in running simulation

Chapter 11 Maximum running speed of T.rex

11.1 A relation of our work with preceding study
11.2 What is the maximum running speed of *T.rex* ?
11.3 A relation between relative stride length and relative velocity
11.4 On the Froude number discussion

Chapter 12   A problem of the position of center of mass

12.1 CM of various posture
12.2 CM of running motion

Chapter 13 Mechanical power for 14.1 m/s running

13.1 Mechanical power calculation
13.2 Mechanical power per kilogram of muscle mass, and comparison with the data of extant animals
13.3 Mechanical power calculation for v=9.8 m/s running motion

Chap 14 Maximal running speed of animals

14.1 Mammalian fast running   — Running speed v.s. log(body mass)—
14.2 Running speed with a measure of body length/s
14.3 Fast running performance of reptile
14.4 Running speed estimation of *T.rex* from the data of extant animals.

Chapter 15 Running ability of Tyrannosauridae and scaling property

15.1 Scaling property of theoretical formula
15.2 Young T.rex could run, and adult *T.rex* could not run?
15.3 Running ability of a family Tyranosauridae

Chapter 16 Coclusion

Appendix Unsolved question on the center of mass of *T.rex*

      Simple calculation of the moment of force for a joint

Preface

In a movie we usually happen to watch such a scene that large *T.rex* chases human. For example, people riding on a car had a narrow escape from *T.rex* in the movie "Jurassic Park". It implies that *T.rex* could run faster than human, then people should ride on a car. To us, such a scenario becomes familiar, however, in science it is not always true. In 2002, a science article was published from the journal "Nature" which is one of the most authoritative science journals, the title of article was "*Tyrannosaurus* was not a fast runner". Since then, the idea seems to be established in science field. Then, it appears to be a gap between science and movie. If the scientific theory is correct, *T.rex* could not run fast. Then, it could not chase any prey. So *T.rex* was supposed to eat dead meet, which is just like today's hyena.

However, the author never believed that *T.rex* was a scavenger, and always ate dead meet. Let's see the picture (cover page of this book), which is the bone of *T.rex* that we often see in a museum. To me it is awful outlooks, and I could not believe that it was a scavenger. It is natural to me to imagine that it was a hunter and chased other animals for hunting. Then, I have been wondering which scenario is true since the publication of the theory. This book is intended to answer this problem. The author accomplished calculation to evaluate *T.rex* running ability based on all available related scientific knowledge. The description of this book is the science research level. So, it may be too be difficult to read for general reader who is not familiar to read science research article.

This book is composed of two parts. Basic biomechanics of the musculoskeletal is introduced in the first half of this book. These descriptions may be helpful to understand theoretical calculation of running ability of bipedal dinosaurs. In the latter half of this book the author presents the result of detailed calculation on the possibility of fast running of *T.rex*.

Before going into describe the content, one point is remarked at this section.

About a name *Tyrannosaurus*

Tyrannosaurus is a name of genius, but only one species is known for the genus. *Tyrannosaurus rex* is a name for the species. *Tyrannosaurus rex* is called as *T.rex* for short, then, hereafter the author call as *T.rex* which is the target of the animal to evaluate running ability in this book. Scientific classification on a family Tyrannosauridae is as follows,

Family: Tyrannosauridae

    Genus; *Albertosaurus*
              Gorgosaurus
              Alioramus
              Daspletosaurus
              Tarbosaurus
              Teratophoneus
              Tyrannosaurus
              Zhuchengtyrannus

    Genus; Tyrannosaurus

        Species *Tyrannosaurus rex*   (*T.rex*)

Running ability of a family Tyrannosauridae is discussed in Chap. 15.

Acknowledgements

The author thanks the late Prof. K. Nagata and Prof. H. Kubotani for continual encouragements. The author is also grateful to my wife Maiko U. and my son Yuri U. for sharing happy time during writing the manuscript. This book is dedicated to the memories of my late mother Natsue U.

Biomechanics for bipedal dinosaurs

How fast *Tyrannosaurus* could run

Chapter 1 Structure of Skeletal Muscle

This chapter describes the basic structure of skeletal muscle. Knowing the structure of muscle is helpful to understand the ability of locomotion of animal, such as dinosaurs.

Skeletal muscles are connected to the bone through a tendon. For example, the gastrocnemius muscle is a large posterior muscle of the calf of the leg. It originates from the back of the thighbone (femur) and kneecap (patella), and is attached to the Achilles tendon at the heel. The gastrocnemius shortening leads to the heel up, and movement with the foot downward. This provides the propelling force of walking and running. Shortening of the gastrocnemius muscle leads to the rotation of ankle joint, however, elastic function of the Achilles tendon plays an important role in movement. This will be discussed in detail in Chap.7.

1 Tendon (bundles of collagen (proteins))

Skeletal muscles are attached to bones by tendon. Tendons are composed of relatively few cells and abundant extracellular matrix. The cells produce the extracellular matrix by metabolism. The extracellular matrix is composed of elongated collagen fibres. Collagen is a group of proteins, which makes up about 30% of the whole body of mammals. Collagen also exists in ligament, skin, bone, cartilage, etc. This extracellular matrix facilitates the property of elasticity of tendon.

Tendon cannot contract, but produce motion together with muscle. Then, it is called "passive" element. On the contrary, muscle is called "active" element, because muscle produces motion by shortening. However, recently it is recognized that elastic property of tendon plays a critical role in movement. Then, the system is called as MTC (muscle tendon complex).

If we consider locomotion of bipedal dinosaur such as *T.rex*, descriptions only on muscles are not sufficient. Considering entire function of the muscle tendon complex (MTC) is crucially important. This issue will be discussed in detail in Chap.7.

2 Bundle of muscle fibres (= muscle fascicle)

The skeletal muscles are composed of many subunits called a bundle of muscle fibres. The muscle fibre is muscle cell, and the fibres are encased in a dense connective tissue called a perimysium. The bundle of muscle fibres is also called as a muscle fascicle.

3 Myocyte = muscle cell = muscle fibre (=fibre)   = myofibre
   [ϕ =10~150 μ m = 10^{-5}~10^{-4} m]

Bundle of muscle fibres is made up of muscle cells. The muscle cell is called mycocyte, and also called muscle fibre. In the united states muscle fibre is spelled as muscle fiber. (Muscle fibre is British usage.)

Chapter 1 Structure of Skeletal Muscle

The muscle fibre is also called myofibre. Then, muscle cell = muscle fibre = muscle fibre = myocyte = myofibre. The diameter of muscle fibre is about 10~150 μ m.

4 Myofibril [ϕ =1 μ m = 10^{-6} m]

The muscle fibre (muscle cell) is composed of hundreds to thousands of tube organelle called myofibrils. The lengths of myofibrils are equal to muscle fibres, and the diameters of those are about 1 μ m = 10^{-6} m.

Myofibrils are made up of chains of repeating units called sarcomeres. Sarcomeres contain two types of protein filaments, myosin and actin.

5 Sarcomere

The myofibril contains repeating sub-units called sarcomeres. The sarcomere is a functional unit of the contractile system in the muscle. An individual sarcomere contains many parallel protein filaments, myosins and actins. Sliding movement of myosins and actins causes contraction of the muscle.

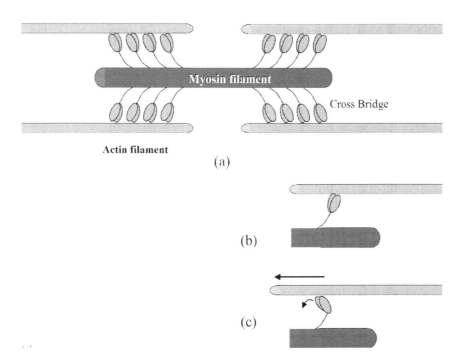

Fig.1.1 Myosin has two heads, and the myosin heads attach to the binding sites on the actin filaments to form cross-bridges (a). By bending the heads, myosins create force to slide the actin filaments just like pulling the rope as (b) and (c).

Chapter 1 Structure of Skeletal Muscle

6 Molecular mechanism of the movement of myosin and actin

Sliding movement of myosin and actin filaments is the fundamental mechanism of muscle contraction. Attaching and detaching of myosin heads from actins causes the filaments to slide relative to one another. This movement reduces the sarcomere length, although filaments length does not change. Then, this is called sliding movement. The mechanism of sliding movement is the followings;

(1) A myosin molecule is made up of two heavy chains. In each chain, there is a head and neck domain.
(2) Myosin head is attached to a place of actin filament.
(3) ATP (nucleotide adenosine triphosphate) attaches to myosin head.
(4) This causes the myosin head to be released from actin filament.
(5) ATP is split to ADP (adenosine diphosphate) and inorganic phosphate Pi by hydrolysis.
   This reaction releases energy.
(6) The energy from the ATP causes the change of angle to the myosin head.
(7) The myosin head binds the actin filament at another site.
(8) Power stroke occurs by this movement, which causes sliding motion of myosin and actin filament.
   The binding of myosin and actine filament is called cross bridge.

7 Length - tension relation

The force of muscle contraction varies as the length at which it is held. The maximum force is produced when the sarcomeres are at resting length (Fig.1.2(a)). In this state overlap of actin and myosin filaments is the maximum in which number of cross-bridges is also maximum.

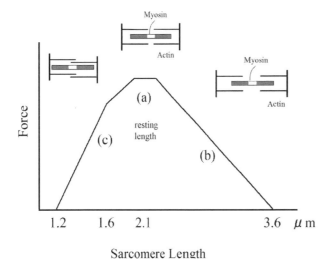

Fig.1.2 Force-length curve of muscle fibre stimulated at different lengths.

The actin filaments have a short bare zone in the middle. Then, the maximum force occurs at not just a single length. This gives a plateau in the force-length relationship, rather than a distinct peak. This makes muscle to generate the maximum force in the wider range of length. If the sarcomeres are lengthened, the force decreases linearly with the length (Fig.1.2(b)). This is because overlap of actin and myosin filaments decreases with the increase of sarcomere length. If the sarcomere shortens to less than resting length, the force decreases steeply rather than the case for lengthening (Fig.1.2(c)). Because, the actin filaments start to interact with the oppositely directed cross-bridge sites.

8 The force of muscle is proportional to its cross-sectional area

The length of overlap of myosin and actin filaments is proportional to the number of cross bridge that myosin heads attach to actin filament. Then, the length of overlap is linearly related to the force that the sarcomere produces (Fig.1.3(a)). However, the forces of neighboring sarcomeres are equal to each other, because those are in serial arrangement (Fig.1.3(b)). It is just like spring that is serially arranged.

On the contrary, the force of the sarcomere in parallel arrangement is multiplied by its number (Fig.1.4). Then, the force of muscle is proportional to its cross-sectional area.

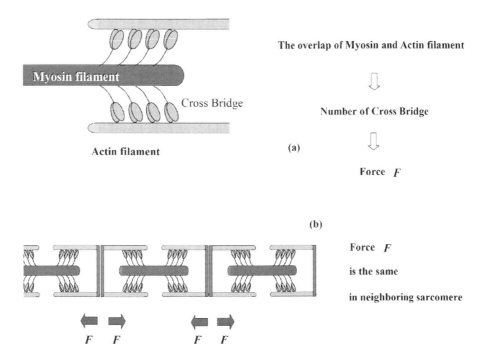

Fig.1.3 (a) The length of overlap region is proportional to the number of cross bridge. It is linearly related to the force that muscle produces. (b) The forces of neighboring sarcomeres are equal to each other. Then, the length of muscle fibres in which sarcomeres are in serial arrangement does not contribute to the strength of the force.

Chapter 1 Structure of Skeletal Muscle

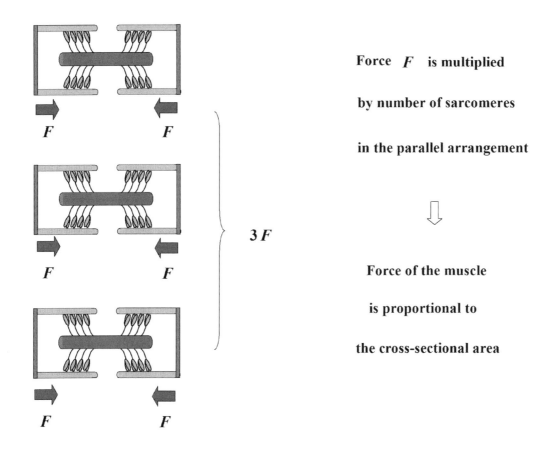

Fig.1.4 Parallel arrangement of sarcomeres brings multiplied force for the muscle.

This idea has been believed over 100 years, i.e. the maximum force per cross-sectional area of muscle is a constant, which is shown in Fig.1.5. However, it is also widely recognized that absolute value of it spans in an extremely wide range as 9~180 N/cm^2. Then, several researchers warned the use of this quantity to evaluate force of muscle from the physiological cross-sectional area.

In 2001, Fukunaga et al. proposed a new idea that muscle volume is an appropriate index to predict muscle strength (Fukunaga et al, 2001; Akagi et al., 2009). The relation is such that torque or moment of force [N · m] is proportional to the muscle volume [m^3]. One reason to propose this index is that the interception at y axis in the graph of force v.s. cross-sectional area is not zero as Fig.1.6(a). The interception at y axis in the graph of torque v.s. volume becomes nearly zero as Fig.1.6(b). This issue will be discussed in detail in Chap.5.

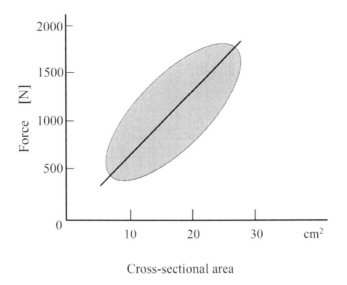

Fig.1.5 The force of muscle is proportional to the cross-sectional area. However, the reported absolute values of the coefficient span in extremely wide range as $9 \sim 180$ N/cm^2, which will be discussed in Chap. 4.

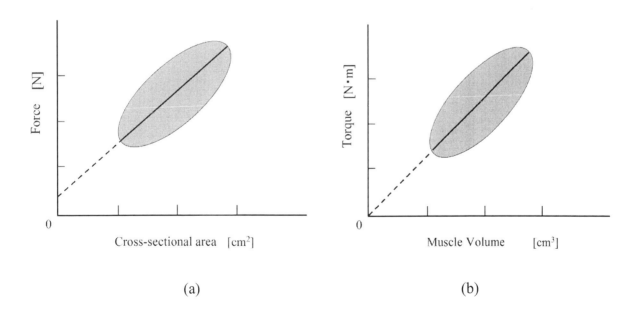

Fig.1.6 In 2002, Fukunaga et al. pointed out that force per cross-sectional area is not an appropriate index, because its interception at y axis is not zero (a). Instead of this, they proposed that torque per

muscle volume is an appropriate index to express muscle strength, because its interception at y axis is nearly zero (b).

References

Akagi, R., Takai, Y., Ohta, M., Kanehisa, H., Kawakami, Y. and Fukunaga, T., Muscle volume compared to cross-sectional area is more appropriate for evaluating muscle strength in young and elderly individuals, Age and Ageing, (2009) 38: 564–569.

Fukunaga. T., Miyatani, M., Tachi, M., Kouzaki, M., Kawakami, Y. and Kanehisa, H., Muscle volume is a major determinant of joint torque in humans, Acta. Physiol. Scand., (2001) 172:249-55.

Chapter 1 Structure of Skeletal Muscle

Chapter 2 Muscle Contraction Property

2.1 Isometric v.s. isotonic and twitch v.s. tetanus

If an electric stimulus is given to dissected bundles of fibres in vitro, the muscle shortens. Such muscle contraction properties are studied in two typical experimental conditions, isometric and isotonic.

- **Isometric** The muscle is in a condition of the same length throughout measurement.
- **Isotonic**   The muscle shortens under a constant load.

Numerous scientific works have been reported discussing on isometric force of muscle in the field of muscle physiology, etc.. However, the following is rarely described. Muscle produces a force by contraction, i.e., by shortening of muscle length. Then, how the muscle generates force in the condition of the same length, namely, isometric condition ?   The answer is that tendon attached to the muscle is stretched according to the shortening of the muscle. Then, a role of tendon in muscle contraction plays a crucially important role. This issue is discussed in Chap.7 .

There are two extremes of electrical stimulation, twitch and tetanus.

- **Twitch** A single stimulus is given to the muscle. Relatively weak force is produced.
- **Tetanus** High frequency stimulation is given. The maximum force is observed.

Twitch
If a single electric pulse is given to muscle, muscle is shortened. The force that muscle generates in shortening is relatively weak (Fig.2.1).

Tetanus
Series of electric pulse stimulus is called tetanus. If tetanus is given, muscle produces several times larger force than twitch force.

When a tetanic stimulus is given on the condition of isometric contraction, the produced force depends on the length of the muscle while contracting. Maximal isometric tension (Po) is produced at the muscle's optimum length, where the length of sarcomeres of muscle is on the plateau of the length-tension curve (Fig. 1.2).

Chapter 2 Muscle Contraction Property

When muscle is under the condition of isometric, i.e., keeping the same length, the observed force is called as "isometric tetanus".

The observed force is divided by cross-sectional area of the muscle, then, the quantity isometric tetanus expresses tension, which is usually described as P_o [N/m²]. Confusingly, P_o [N/m²] is sometimes called force in the literature, although, it expresses a stress, strictly speaking.

As an example of experimental result, Milligan et al.'s work published in 1997 is introduced. In the experiment tissue in the dimensions of length 7.1-12.4mm , width 5mm and thickness 0.1-0.2mm are dissected from the ventral mantle of cuttlefish.

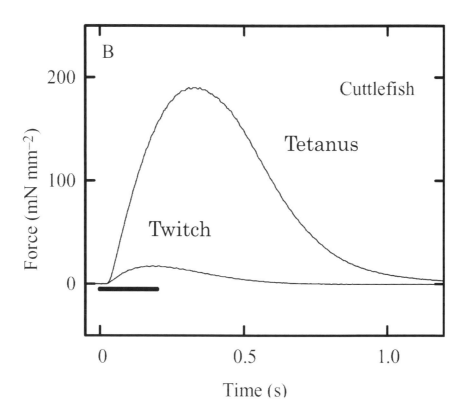

Fig.2.1 Superimposed records of twitch and tetanic force. A tetanus of duration 0.2 s is given at Time=0 s for cuttlefish, that is shown as heavy horizontal bar. (This figure is reproduced with permission from Fig.8 of The Journal of Experimental Biology, Milligan et al., 1997, vol.200, pp2425-36).

When a single pulse of electric stimulus is given, the muscle produces a relatively weak force. This is called twitch force shown as "twitch" in Fig.2.1 . On the other hand, when a series of stimulus is given, isometric muscle force raises several times larger than twitch as shown in Fig.2.1

Chapter 2 Muscle Contraction Property

2.2 P_o [N/m²] "Isometric tetanus" = "Maximum isometric tension"
= "Isometric tetanic tension"

If stimulus strength and stimulus frequency are increased, the isometric force rises. However, there is the maximum force that muscle can produce. The force is called maximum isometric tetanus, which is usually expressed as P_o [N/m²]. This quantity is also called isometric tetanic tension, and also maximum isometric tension. Fig.2.2 shows stimulus frequency dependence of the isometric tetanus force. The pulse duration is 0.2 second, which is shown as thick horizontal bar in Fig.2.2. In the duration 5~150 Hz electric pulse is involved. We observe the rises of the maximum of the force in Fgi.2.2. However, the maximum force does not exceed a certain limit.

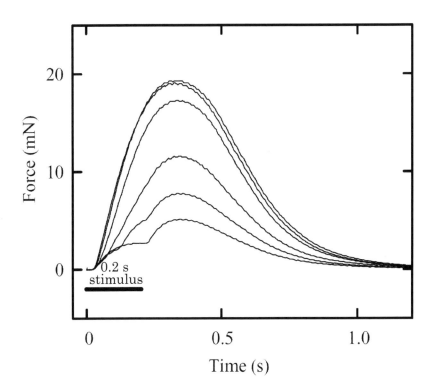

Fig.2-2 Stimulus frequency dependence of the isometric tetanic force. The superimposed data show the force in response to stimulation at frequencies 5, 150,20,50,100 and 150 Hz. The stimulus duration is 0.2 seconds shown as the thick horizontal bar. (Reproduced with permission from Fig.7 of The Journal of Experimental Biology, Milligan et al., 1997, vol.200, pp2425-36.)

Chapter 2   Muscle Contraction Property

2.3   Hill's equation: A basic model expressing muscle contraction:

Experimentally, an inverse relationship between load (force) and velocity of shortening was known. In 1938, A. V. Hill derived basic equation to express the relations, and showed that experimental data are fitted into the equation. It is now known as Hill's equation. In this subsection, mathematical property of this equation is introduced.

Concerning to the mechanical work of muscle, two quantities, work and heat play important roles. A product of the force F and shortening distance x is the work $W = F \cdot x$. Energy supplied by ATP is consumed in a muscle, and transferred into heat. Even in a condition that shortening distance is zero ($x = 0$), but with a load, the muscle generates heat. And for the basic characteristic of muscle shortening, the followings are said using currently used terminology.

1   For slow shortening of muscle, relatively strong force is produced by muscle. On the contrary, for fast shortening of muscle weak force is produced by muscle. (This implies hyperbolic function of force F and velocity v of shortening.)

- large W (work) for slow v (velocity).
- Weak F (force) for fast v.
- These lead to hyperbolic relation of F and v.

2   Muscle does not exert any work when a shortening distance is zero, (x=0). In this condition the maximum force is generated by muscle.   Muscle does not also exert any work in a case that muscle shortening occurs with zero force (F=0).

- W=0 (work is zero)   when x=0   in $W=F \cdot x$
- W=0 (work is zero)   when F=0   in $W=F \cdot x$

In 1938's paper Hill presented the following derivation:

In a case of no load is applied, the observed heat per unit distance shortening (1 [cm]) was constant in the experiment. Then, shortening heat is proportional to shortening a distance x. Let us describe the heat as ax, where a is a constant having a dimension of force. When a force F is applied, the work of muscle for shortening distance x is Fx. Then, total work that the muscle consumes becomes $W = (F + a)x$. In a case that the force F is a constant, the rate of energy liberation (i.e. power P) becomes $P = (F + a)v$, where v is muscle shortening speed (velocity) defined by. $v = dx/dt$.

- Work $W = (F + a)x$
- Power $P = (F + a)v$

Chapter 2 Muscle Contraction Property

In addition, Hill stated the following. Let F_0 be the full isometric force, a/F_0 becomes a constant. He wrote that the mean value of a/F_0 was about 0.25 in eleven experiments.

- $$\frac{a}{F_0} = \frac{1}{4}$$

The ax is heat when no load is applied, where a is a constant, and x is shortening distance. The F_0 is the full isometric force.

Further, Hill found linear relation of the energy liberation rate (power P) and the force F. The energy liberation rate (power P) diminishes with the increase of F, and the rate becomes zero when the load is the maximum F_0. Then, the relation of the rate (power P) becomes,

$$P = (F+a)v = b(F_0 - F) \quad , \tag{2-1}$$

where b is a constant defining the absolute rate of energy liberation. This equation can be rewritten as Hill's equation.

- **Hill's equation**

$$(F+a)(v+b) = (F_0 + a)b = \text{const.} \tag{2-2}$$

Eq.2.2 is famous Hill's equation which describe the relation of applied force F and shortening velocity v. This is the form of hyperbolic function with asymptotes $v = -a$ and $F = -b$. In a F-v graph F_0 is the intercept on the F axis ($v=0$), and $F_0 \frac{b}{a}$ is the intercept of v axis ($F=0$) as shown in Fig.2.3.

Hill's equation has two extremes:

- Force F has the maximum when $v=0$.
- Velocity v has the maximum when $F=0$.

These characteristics tell that muscle generates large force for slow shortening, and shows the maximum when shortening speed is zero ($v=0$). And, muscle shortening speed v becomes large when force decreases. The maximum shortening speed is achieved when no force is applied to muscle. Hill's eq. is the simple and beautiful form, and very much useful to understand numerous experimental results of muscle shortening.

Chapter 2 Muscle Contraction Property

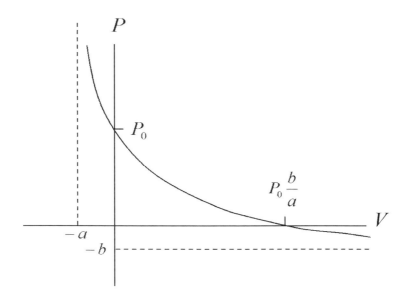

Fig.2.3 Hill's equation of force F and velocity v for muscle contraction.

2.4 Mathematical properties of Hill's equation

Eq.2.2 is solved for v as,

$$v = b\left[\frac{F_0 - F}{F + a}\right] . \tag{2.3}$$

If we multiply F for both sides, we obtain the expression for power P of the muscle,

$$P = Fv = Fb\left[\frac{F_0 - F}{F + a}\right] . \tag{2-4}$$

The power Fv becomes zero when $F=0$ and $F=F_0$. The power becomes the maximum when $F = -a \pm \sqrt{(F_0 + a)a}$, which is obtained by the following calculation.

Chapter 2 Muscle Contraction Property

$$\frac{d}{dF}\left[Fb\left(\frac{F_0 - F}{F + a}\right)\right] = 0, \qquad (2.5)$$

$$b\frac{F_0 - 2F}{F + a} - Fb\frac{F_0 - F}{(F + a)^2} = 0 \qquad (2.6a)$$

$$(F_0 - 2F)(F + a) + F(F - F_0) = 0 \qquad (2.6b)$$

$$F^2 + 2aF - aF_0 = 0 \qquad (2.6c)$$

$$F = -a \pm \sqrt{(F_0 + a)a} \qquad (2.6d)$$

For example, let each constant be the unit as $a=1$, $b=1$, $F_0=1$. In this case, P is the maximum at $F = -1 \pm \sqrt{2} = 0.414$. Figure 2.4 shows the relation of P and F for this case.

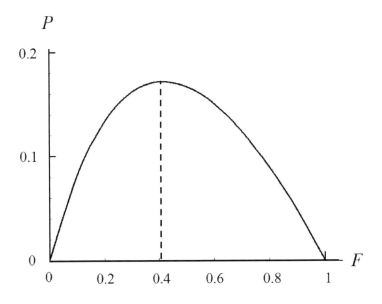

Fig. 2.4 A relation of power P and force F derived from Hill's equation. In the case that all the coefficients are unity, i.e., $a=1$, $b=1$ and $F_0=1$, the maximum of P is observed at $F = -a \pm \sqrt{(F_0 + a)a} = -1 \pm \sqrt{2} = 0.414$.

We observe that power P is zero at $F=0$ and $F=1$ ($=F_0$). And, the maximum of P is located at almost the middle point of $F=0$ and $F=1$ ($=F_0$).

Chapter 2   Muscle Contraction Property

Eq.2.5 can be calculated into,

$$Fv = -bF - \frac{1}{F+a}ab(F_0 + a) + b(F_0 + a) \tag{2.7}$$

Then, the power Fv becomes the function as,

Power $\quad P = Fv = -c_1 F - \dfrac{c_3}{F+c_2} + c_4 \quad,$ (2.8)

where c_i is constant for $i = 1 \sim 4$. An actual experimental data of power will be shown later.

Inversely, force F is written as the following from Hill's equation.

$$F = \frac{F_0 b - av}{v+b} \quad, \tag{2.9}$$

Then, power is obtained by the product with v.

$$P = Fv = \frac{F_0 b - av}{v+b} v \quad, \tag{2.10}$$

The maximum of P is located at the point,

$$\frac{dP}{dv} = \frac{d}{dv}\left(\frac{F_0 b - av}{v+b} v\right) = 0 \tag{2.11}$$

Then, we obtain for $\dfrac{dP}{dv} = 0$ at the following.

$$v = -b + b\sqrt{1 + F_0/a} \quad, \tag{2.12}$$

If we assume all the coefficients are zero, namely, $a = b = F_0 = 0$, $v = -1 + \sqrt{2} = 0.414$. The graph of Eq.(2.10) for this condition yields the same of Fig.2.4 except for horizontal axis being v instead of F.

Chapter 2 Muscle Contraction Property

This is because that $P(F)$ and $P(v)$ has the same functional form as,

$$P(F) = \frac{F_0 - F}{F + a} Fb \quad , \quad P(v) = \frac{F_0 b - av}{v + b} v \tag{2.13}$$

The origin of this property comes from symmetrical structure on F and v in Hill's equation, $(F + a)(v + b) = \text{const}$.

Notice

F is force [N], and P is power [J/s], but sometimes P is used for stress [N/m²]

It is noted that P stands for power [J/s] in physical textbook. When force [N] is divided by cross sectional-area A [m²], the resulting quantity is stress [N/m²]. But, in the field of muscle physiology stress is expressed by P [N/m²]. Then, P_o expresses the maximum tension in the unit of [N/m²]. Confusingly, this quantity stress is sometimes called as force in the literature of muscle physiology.

P stands for power [J/s] in physical textbook.
P stands for stress [N/ m²] in muscle physiology.
   Force divided by cross sectional area
   (but, sometimes called as force.)

The force F in Hill's equation Eq.(2.2) is often replaced by stress P in the literature as,

$$(P + a)(v + b) = \text{const} \quad , \tag{2.14}$$

where P stands for stress, but sometimes called as force.

Interestingly, Hill's relation Eq.(2-2) can be applied for the experiments of bundles of muscle fibres, and also for the ones of arm and leg which are studied in sports biomechanics. The later case will be discussed at the end of this chapter. In the next subsection the author introduces the other relations proposed so far as improved version of Hill's relation.

Chapter 2 Muscle Contraction Property

2.5 Improved version of Hill's relation.

As for improved version of Hill's equation, Edman's equation was proposed in 1976 (Edman et al., 1976), and HYP-LIN was proposed by Marsh and Bennett in 1986 (Marsh and Bennett, 1986),

(1) $\quad v(P) = b\left[\dfrac{P_0 - P}{P + a}\right]$, Hill's equation, (2.15)

(2) $\quad v(P) = b\left[\dfrac{P_0 - P}{P + a}\left\{1 - \dfrac{1}{1 + \exp(-k_1(P + k_2))}\right\}\right]$, Edman's equation, (2.16)

(3) $\quad v(P) = \dfrac{B(1 - P/P_0)}{A + P/P_0} + C(1 - P/P_0)$, HYP-LIN (hyperbolic-linear). (2.17)

The right-hand side of Eq.(2.15) and the first term in the right-hand side of Eq.(2.16) are identical to the right-hand side of Hill's equation (2.2). The second term in the right-hand side of Eq. (2.16) and Eq.(2.17) is a correction term to the original Hill's equation. P_0 stands for the maximum isometric stress. The others except for v and P are constant variables.

Scale of variables

Usually, each variable is scaled by its maximum value or typical value. The scaled variable becomes dimensionless. For example, let us define dimensionless stress as $\hat{P} = P/P_{tw}$, where P_{tw} is the peak twitch stress. Then, Hill's equation becomes,

$$V(P) = b\left(\dfrac{\dfrac{P_0}{P_{tw}} - \hat{P}}{\hat{P} + \dfrac{a}{P_{tw}}}\right). \tag{2.18}$$

Chapter 2 Muscle Contraction Property

In Hill's Eq.(2.2) the parameter a have dimension of stress P, and the b has dimension of velocity v. (Hereafter, veclocity v is expressed as V, as in the literature.) Then, let us define $b = \dfrac{aV_{max}}{P_0}$. The V_{max} becomes the intercept on the velocity axis. And, let us define $P^* = \dfrac{P_o}{P_{tw}}$. The P^* becomes the intercept on the scaled stress axis. Eq.(2.18) becomes

$$V(P) = \frac{aV_{max}}{P_0}\left(\frac{P^* - \hat{P}}{\hat{P} + \dfrac{a}{P_{tw}}}\right)$$

$$= \frac{aV_{max}}{P_0}\frac{1}{\dfrac{a}{P_{tw}}}\left(\frac{P^* - \hat{P}}{\dfrac{P_{tw}}{a}\hat{P}+1}\right) = V_{max}\frac{P_{tw}}{P_0}\left(\frac{P^* - \hat{P}}{G\hat{P}+1}\right) = \frac{V_{max}}{P^*}\left(\frac{P^* - \hat{P}}{G\hat{P}+1}\right) \qquad (2.19)$$

Then, we have,

$$V(\hat{P}) = \frac{V_{max}}{P^*}\left(\frac{P^* - \hat{P}}{G\hat{P}+1}\right) \qquad (2.20)$$

The P^* is the intercept on the scaled stress axis $\hat{P}$. In some work velocity V is scaled by the length L_{tw} at which peak isometric twitch force is produced. Figure 2.5 is a such example accomplished by Milligan et al. (Milligan et al. 1997).

In some work velocity V is scaled by V_{max} as $\hat{V} = \dfrac{V}{V_{max}}$. In such a case, both variables become dimensionless as,

$$\hat{V}(\hat{P}) = \frac{1}{P^*}\left(\frac{P^* - \hat{P}}{G\hat{P}+1}\right) \qquad (2.21)$$

Chapter 2 Muscle Contraction Property

Each variable in Eq.(2.21) are the followings.

P_{tw} [N/cm²] The peak twitch stress. (Sometimes, this is called "force").
P_0 [N/cm²] The maximum isometric stress (Sometimes, this is called "force").
V_{max} [m/s] The velocity at power being zero ($P=0$).
L_0 [m] The length at a rest.
L_{tw} [m] The length at which peak isometric twitch force is produced.

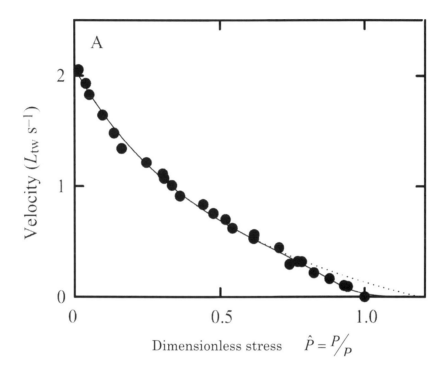

Fig.2.5   Stress - velocity relationship for a single squid preparation. The broken line expresses was fitted curve to the data shown as the black circle below using Hill's equation. The solid line shows fitted curve using Edman's euation. Note that stress P [N/cm²] is often called force in he literature. (Reproduced with permission from Fig.10 of The Journal of Experimental Biology, Milligan et al., 1997, vol.200, pp2425-36.)

Stress - velocity relationship for a single squid preparation is shown in Fig.2.5. Stress [N/cm²] is often called as force in the literature. Then, this graph is called force - velocity relationship in the reference (Milligan, et al. 1997). The black circle shows experimental data, and the solid line expresses fitted result using Edman's equation. The broken line for $V<0.78$ shows fitted line using Hill's equation. From this figure, we observe that force - velocity relation shows hyperbolic relation, and Hill's equation well describes the experimental data as the first approximation. This relation shows the muscle characteristic that muscle shortening velocity is high for low load, and velocity is low for high load.

Chapter 2 Muscle Contraction Property

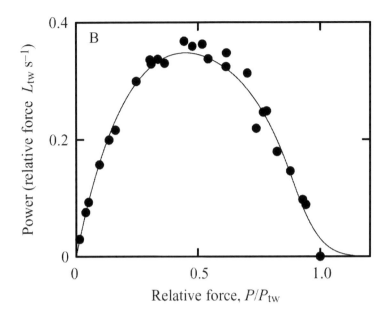

Dimensionless stress $\hat{P} = P/P$

Fig. 2.6. Stress - power relationship corresponding the data of Fig2.5. Stress is called as force in this reference, then this graph is called force-power relationship. (Reproduced with permission from Fig.10 of The Journal of Experimental Biology, Milligan et al., 1997, vol.200, pp2425-36.)

Fig. 2.6 shows the mechanical power output calculating from the data of Fig.2.5. Maximum power was produced at $\hat{P} = 0.45 \pm 0.03 P_{tw}$ (mean ± S.E.M., N = 8).

Isometric mechanical property is listed in Table 2.1.

	Squid (N=6)	Cuttlefish (N=7)
P_o (N/cm^2)	26.2±1.6	22.6±1.9
P_{tw}/P_o	0.18±0.01	0.10±0.02

Table 2.1

Talbe 2.1. Isometric mechanical properties of thin slice mantle preparations of the squid and cuttle fish. P_o is the isometric tetanus, i.e., peak force in a tetanus of 0.2s stimulation at 100-150 Hz. The P_{tw}/P_0 expresses the peak twitch-tetanus ratio.

Table 2-1 shows obtained data of the maximum isometric tension P_o and P_{tw}/P_0 of thin slice mantle preparations of squid and cuttle fish. Values of the isometric tetanic tension P_o show typical value of this quantity $15 < P_0 < 30 \ [\text{N}/\text{cm}^2]$.

2.5 Hill's equation is also used in sports science

Hill's equation was intended to explain muscle shortening property published in 1955. At that time, molecular mechanism of muscle shortening was not known. Muscle contraction mechanism in which myosin and actin achieve sliding movement was revealed by the development of molecular biology in 1980's. As is introduced in the previous sections, Hill's equation has been well valid for explaining contraction property of muscle fibres in vitro.

Interestingly, Hill's equation has been also used in sports science widely. In the study, Hill's relation of *F-V* has been used for the explanation of the relation of force and velocity of muscle in the motion of human skeletal muscles. For example, those are bending motion of knee joint and/or elbow joint, etc. Figure 2.7 shows a result of *F-V* curve for the adductor pollicis muscle (Jones, 2010: from de Ruiter et al., 1999). Strictly, the horizontal axis represents angular velocity, then, it should be denoted as ω [degree/second]. The unit [degree/second] can be translated into the unit [1/second]. Note that velocity *V* [m/s] is obtained by the product of arm length *r* [m] and ω [1/s], i.e., $V = r \cdot \omega$. The data of mechanical power is also shown in Fig.2.7. This graph was drawn to study fatigue effect of muscle. Figure 2.7 reveals that both of force and power decreases caused by fatigue of muscle.

Chapter 2 Muscle Contraction Property

- *Force-Velocity* measurement at International Space Station -

An example is shown as Fig.2.7 (Trappe, s. et al., 2009). The data was taken in the study of the influence of long stay at ISS (International space station). The hyperbolic relation between *F* and *V* as in Hill's equation is observed. During 6 months stay at ISS, seven crews (subjects) accomplished 2~4 hours physical exercise per day. However, 20-29% decline of force across the velocity spectrum was observed. The other data in the study also shows clearly the reduction of muscle ability in spite of physical training at ISS during the stay.

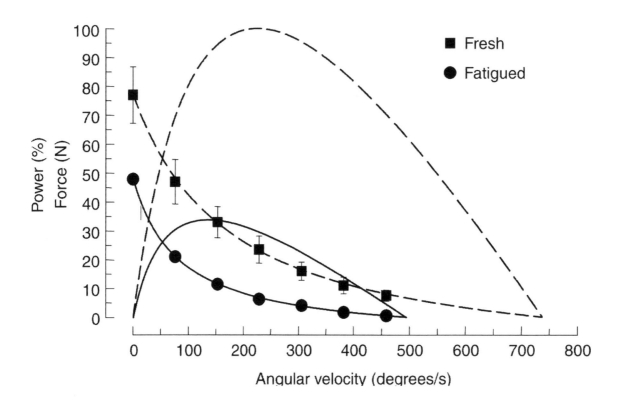

Fig. 2.7 Force-velocity relationships of fresh and fatigued human adductor pollicis muscle, together with power (Jones, 2010: from de Ruiter et al., 1999).

References

De Ruiter, C. J. Jones, D. A. Sargeant, A. J., and De Haan, A., Themeasurement of force–velocity relationships of fresh and fatigued human adductor pollicis muscle., Eur. J. Appl. Physiol. (1999) 80: 386–393.

Edman, K. A. P., Mulieri, L. A., and Scubon-Mulieri, B., Non-hyperbolic force-velocity relationship in single muscle fibres. Acta physiol. scand. (1976) 98: 143-156.

Hill, A.V., The heat of shortening and dynamics constants of muscles. Proc. R. Soc. Lond. B. (1938) 126; 136–195.

Jones, D. A., Changes in the force–velocity relationship of fatigued muscle: implications for power production and possible causes, J. Physiol. (2010) 588:2977-286:　Published online 2010 June 14. doi:　10.1113/jphysiol.2010.190934

Marsh, R. L., and Bennett, A. F., Thermal dependence of sprint performance of the lizard Sceloporus occidentalis, J. Exp. Biol. (1986) 126: 79-87.

Milligan, J. B. and Curtin, N. A., Contractile Properties of Obliquely Striated Muscle From the Matle of Squid (Alloteuthis subulata) and Cuttlefish (Speia officinalis), J. Exp. Biol. (1997) 200: 2425-2436.

Trappe, s. et al., Exercise In Space: Human Skeletal Muscle After 6 Months Aboard The International Space Station, Appl. Physiol. (2009) 106:1159-68.

Chapter 3   The maximum tetanic tension P_0.

Chapter 3   The maximum tetanic tension P_0.

3.1 The maximum muscle stress σ, and the maximum tetanic tension P_0

The force of muscle is proportional to the cross-sectional area of it. This idea was proposed by Weber in 1856[1]. The quantity is called maximum muscle stress, and it is denoted as σ. After that, vast number of studies has been published on this issue over hundred years.   Confusingly, there is another quantity called the maximum tetanic tension usually denoted as P_0. Both σ and P_0 are similar quantity, i.e., force per cross-sectional area, then both dimension is the same [N/m²]. However, the usage and meaning of it is not always the same as schematically shown in Fig.3.1. The measurement of σ has been accomplished by calculating the ratio of force per cross-sectional area of arm, leg or trunk muscle. An order of the cross-sectional area of it is around 10 cm². The experimental results have been published for 150 years in 1850's~2000's.

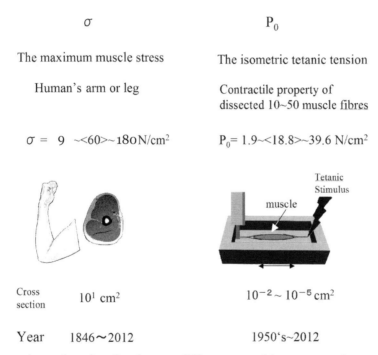

Fig. 3.1. Schematic explanation for the two different quantities: σ maximum muscle stress and P_0 isometric tetanic tension. About the maximum muscle stress the known minimum and maximum for σ are 9 N/cm² and 180 N/cm², respectively. The average and the standard deviation is 60±34 N/cm². Characteristic area for it is 1×10^1 cm². On the isometric tetanic tension P_0, observed value of it ranges as 1.9~39.6 N/cm². The average and standard deviation is 18.8±9.2 N/cm² over 56 references listed in Table 3.1, except for molluscs and crustaceans. This value depends on the choice of literature, however,

Chapter 3　The maximum tetanic tension P_0.

the range agrees with the consensus of the value of P_0, i.e., $9 < P_0 < 27$ N/cm^2 in this research area. Characteristic cross-sectional area for P_0 is small as 1×10^{-2}~1×10^{-5} cm^2.

The P_0 is the maximum isometric force, which has been used in the study of contractile property of muscle fibres. In most studies, small bundles of 10~50 muscle fibres are dissected, and contractile properties are studied *in vitro,* or in some cases *in vivo*. The cross section is so small to measure directly, then, it is calculated by the mass divided by its length and density. Typical range of cross-sectional area for determining P_0 is 1×10^{-2} ~ 1×10^{-5} cm² [2-9]. Note that few experiments reported for it as in the range of 1×10^{0}~1×10^{-2} cm² [10-12]. In 1980's~1990's a considerable amount of data has been collected based on this methodology.

Thus, the maximum muscle stress σ and the maximum tetanic tension is a different quantity. The measurements of the maximum muscle stress σ will be introduced in Chap 4. In this chapter, measurements of the maximum tetanic tension are summarized.

3.2　The isometric tetanic tension　P_0

From 1980's contractile property of muscle fibres has been studied actively in biological science. Modern technique using laser and force transducer made it possible to observe less than millimeter and millisecond order measurement. In usual experimental setup bundles of 10~50 muscle fibres of the length 1 mm ~1 cm are dissected from animal, and tetanic stimulation of 50~400 Hz is given in isometric condition. The experimental setup is schematically shown in Fig.3.1. Muscle cross-sectional area is approximated by the mass divided by its length and density. The value of the cross section is small as 1×10^{-2}~1×10^{-5} cm². A quantity of force per cross-sectional area of muscle is called as maximum isometric tension or isometric tetanus in this area. In 1980's and 1990's numerous works have been done to determine this quantity. Reported values of P_0 are summarized in Fig.3.2 and Table 3.1.

Chapter 3   The maximum tetanic tension P_0.

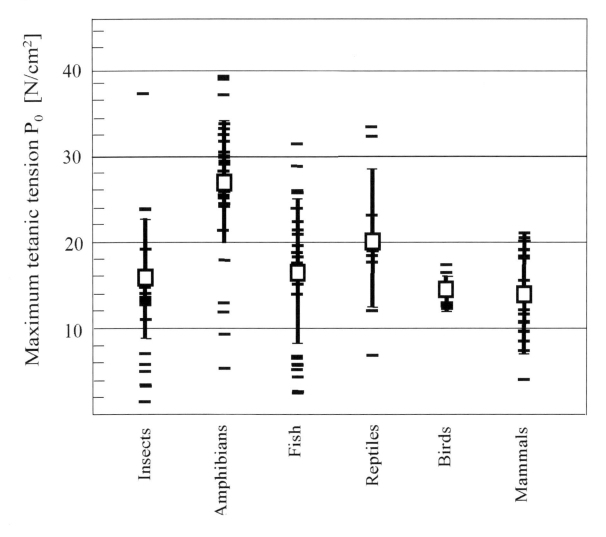

Fig.3.2   69 reference data of the maximum tetanic tension P_0 are shown graphically. Corresponding data set is listed in Table 3.1. A mean value in each published experimental work is expressed as short horizontal bar. White square represents an average in each group, and vertical line expresses standard deviation (SD). Each data usually contains standard deviation (SD) or standard error of the mean (SEM) as in Table 3.1. However, the data of SD and SEM are not shown, because it makes the graph too complex.

In Fig.3.2 short horizontal bar expresses reported mean value of P_0 in each experimental work. The white square represents an average in each group, and vertical line expresses the standard deviation (SD). The P_0 of crustacean and molluscs is known to be exceptionally large compared to the other group as in Fig.3.3. As is observed in Fig.3.2, it is hard to settle single value for P_0. It is because that species, fibre type, condition such as temperature and the other factors lead to a different value of P_0. The lowest and the highest are 1.9 ± 0.8 (SD) N/cm² (Ref. 27), and $39.6 \pm$

Chapter 3 The maximum tetanic tension P_0.

5.4 (SD) N/cm² (Ref. 49), respectively, except for crustacean and molluscs. An average of overall species except for crustacean and molluscs is 18.8 ±9.2 (SD) N/cm². Although this value depends on the choice of literature, the value of P_0 is regarded as 9~27 N/cm² in this area, which coincides with our statistical result 18.8 ±9.2 (SD) N/cm². Note that the values in the range of 30~40 N/cm² is considered to be large in this research field, however, those have been reported frequently in the literature as seen in Fig.3.2.

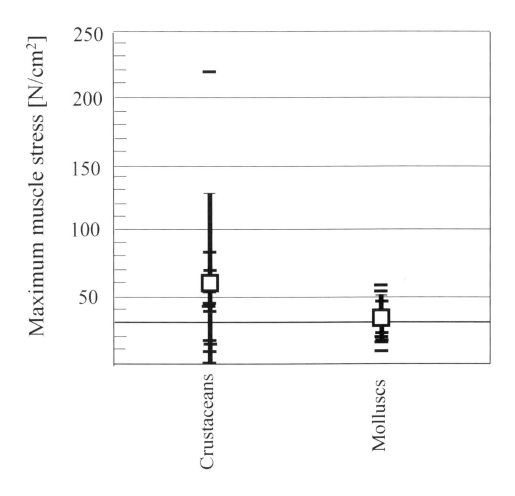

Fig.3.3 The data of the maximum tetanic tension P_0 for crustaceans and molluscs.
Large value of P_0 has known for these groups.

Chapter 3　The maximum tetanic tension P_0.

3.3　Running ability of *T.rex* and P_0

The purpose of writing this book is to investigate running ability of *T.rex*. A detailed discussion is given in Chap. 9~16. However, comments concerning to P_0 with this issue are briefly presented in this section.

In the works [14,19,20], which stated inability of *T.rex* running, the muscle stress is expressed using the maximum muscle stress σ. However, its actual value is replaced by the maximum tetanic tension P_0. As seen in this and next chapter, these two are different quantities. Then, the following points are remarked for evaluating true *T.rex* running ability.

First;
The use of maximum isometric tension P_0 is doubtful for evaluating 6071.82 kg *T.rex* locomotion. Experimentally measured cross-sectional area of P_0 of extant animal is in a order of $1 \times 10^{-2} \sim 1 \times 10^{-5}$ cm^2, whilst cross-sectional area of *T.rex* is in a order of 10^3. Then, scale difference of the cross-sectional area of the two is an order of $10^5 \sim 10^8$, which is extremely large. On the contrary experimentally measured cross-sectional area of σ for human is in an order of 10^1. Then, the scale difference of cross-sectional area of σ between human and *T.rex* is an order of 10^1, which is very small. Then, using σ (maximum muscle stress described in the next chapter) is more suitable than using P_0 (maximum isometric tension described in this chapter) for evaluating *T.rex* running ability.

Second;
If we use P_0 instead of σ, a value of 30.0 N/cm² is employed for the maximum tetanic tension P_0 in the works [14,19,20]. However, it is not a consensus of the maximum nor the mean value of P_0. Apparently from Fig.3.2, it is extremely difficult to choose a single value for P_0 for the evaluation of *T.rex* locomotion. For conservative evaluation, P_0=39 N/cm² should be used. Because, two different groups reported the value as 39.6±5.4 (SD) N/cm² (Ref.49), and 39 N/cm² (Ref.55).

Third;
Hutchinson et al's value of σ =30.0 N/cm² in the works [14,19,20] is not accurate citation of the reference. In the papers, a value of σ =30.0 N/cm² is employed by the citation of the reference [15]. However, the reference [15] is an issue concerning to a summary of past published experimental data on P_0. And, a value of 30.0 N/cm² came from the experimental report [16]. In the reference, it is noted that 32.5±1.5 N/cm² and 33.0±2.1 N/cm² were observed for red iliofibularis and white iliofibularis of a desert iguana at 40℃. More precisely, P_0 varied from 25 to 37 N/cm² with the change of temperature of 30℃~40℃, which is read from Fig.3 of the reference [16]. These are shown for the data of reptiles with asterisk in Fig.3.4, which excess 30 N/cm².

Chapter 3   The maximum tetanic tension P_0.

Fourth;

In most of experimental reports, P_0 contains standard deviation or standard error of the mean as in Table 3.1. For example, the highest value 39.6 N/cm² contains the standard deviation of ±5.4 N/cm². This example is shown in Fig.3.4 as a vertical line after arrow at the top of amphibians column. Then, situation is so complicated. Which selection is appropriate for P_0 ? Is it 18.8 N/cm², 33 N/cm², 37 N/cm² , 39 N/cm² or 45 (=39.6+5.4) N/cm². The evaluation of running possibility of *T.rex* would be changed by the choice. The value of 30.0 N/cm² in the works [14,19,20] is not said as an appropriate choice for P_0 according to 69 published experimental results. It does not have any special meaning such as the maximum nor the typical value of P_0

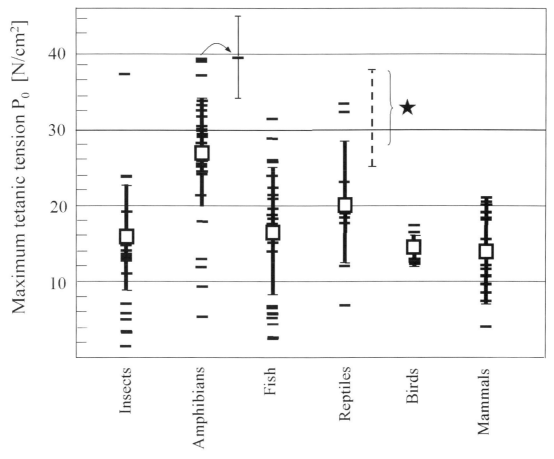

Fig.3.4 Hutchinson et al.'s works employed a value of 30 N/cm² for P_0 in the reference [14,19,20] by citing the work [15]. However, the value in it [15] is a citation of the previous work [16]. The value is read from Fig.3 of it [16] as 25 to 37 N/cm² with the change of temperature of 30℃～40℃, which is shown as solid broken line with asterisk in the column of reptiles. The other many experimental data show larger value than it.   Lännergren reported 39.6±5.4 (SD, *n*=8) N/cm² as shown at the top of amphibians column [49].

Chapter 3   The maximum tetanic tension P_o.

Table 3.1

Common Name	Species	Muscle	P_0 [N/cm^2]			Author	Year	Refs. No.
Insects								
Moth	*Operophtera brucruceat*	flight	13.9			Marden	1995	21
Katydid	*Neoconocephalus triops*	metathoracic	12.6 ±1.3	(SEM)	(n=6)	Josephson	1984	22
		mesothoracic	5.8 ±0.5	(SEM)	(n=6)	Josephson	1984	22
	Neoconocephalus robust	metathoracic	13.7 ±1.8	(SEM)	(n=5)	Josephson	1984	22
		mesothoracic	4.8 ±0.5	(SEM)	(n=7)	Josephson	1984	22
Dragonfly	*Livellula pulcella*	flight	12			Fitzhugh et al.	1997	23
		basalar	11.0			Marden et al.	2001	24
Bumble bee	*Bombus terrestris*	flight	3.7			Josephson et al	1997	25
Locust	*Schistocera americana*	flight	36.3 ±1.4	(SD)	(n=5)	Malamud et al.	1991	26
Beetle	*Cotinus mutalilis*	metathoracic	1.9 ±0.8	(SD)	(n=7)	Josephson et al	2000	27
Hawkmoth	*Manduca sexta*	flight	7.0 ±0.8	(SD)		Marden	1995	28
	Operophtera bruceata	flight	13.9 ±1.0	(SD)		Marden	1995	28
Cockroach	*Blaberus discoidalis*	metathoracic	15.7 ±3.2	(SD)	(n=6)	Full et al.	1998	29
		mesothoracic	19.1 ±2.7	(SD)	(n=6)	Rome et al.	1992	30
			24.0			Josephson	1984	31
Fish								
Sculpin	*Myoxocephalis scorpii*	anterior white	15.9 ±8.2	(SEM)	(n=6)	James et al.	1998	32
		posterior white	16.1 ±5.2	(SEM)	(n=6)	James et al.	1998	32
Antarctic cod	*Notothenia coriiceps*	fast	18.5 ±1.53	(SEM)	(n=9)	Franklin et al.	1997	33
Blue crevally	*Carangus melampygus*	red	4.3			Johnston et al.	1984	34
		white	18.3			Johnston et al.	1984	34
Toadfish	*Opsanus tau*	red	21.4			Rome et al.	1996	35
		white	22.8			Rome et al.	1996	35
		swim bladder	5.6			Rome et al.	1996	35
Antarctic fish	*Notothenia neglecta*	white	22.5			Johnston et al.	1984	34
Scup	*Stenotomus chrysops*	red	19.7 ±2.0	(SEM)	(n=9)	Rome et al.	1992	30
		pink	15.1 ±0.9	(SEM)	(n=22)	Coughlin et al.	1996	36
Grey mullet	*Mugil cephalus*	white	21.0 ±2.8	(SEM)	(n=8)	Johnston et al.	1984	34
Skipjack tuna	*Katsuwonus pelamis*	red	2.4 ±0.2	(SEM)	(n=6)	Johnston et al.	1984	34
		white	15.7 ±1.3	(SEM)	(n=11)	Johnston et al.	1984	34
Kawakawa	*Euthynuus affinis*	red	2.5 ±0.3	(SEM)	(n=10)	Johnston et al.	1984	34
		white	18.8 ±0.8	(SEM)	(n=6)	Johnston et al.	1984	34
Blue marlin	*Makaira migricans*	red	5.7 ±0.9	(SEM)	(n=11)	Johnston et al.	1984	37
		white	17.6 ±2.1	(SEM)	(n=13)	Johnston et al.	1984	37
Antarctic fish	*Chaenocephalus aceratu*	white	21.4 ±1.4	(SEM)	(n=14)	Johnston et al.	1985	38
		red	6.6 ±0.7	(SEM)	(n=9)	Johnston et al.	1985	38
	Notothenia rossi	white	26.0 ±2.2	(SEM)	(n=9)	Johnston et al.	1985	38
		red	6.7 ±0.8	(SEM)	(n=6)	Johnston et al.	1985	38
	Trematomus hansoni	white	26.1 ±2.1	(SEM)	(n=12)	Johnston et al.	1985	38
Dog fish	*Scyliorhinus canicula*	white	24.1 ±2.2	(SEM)	(n=7)	Curtin et al.	1988	39
Shorthorn sculpi	*Myoxocephalus scorpius*	fast myotomal	31.5			Langfeld et al.	1989	40
Dog fish	*Scyliorhinus canicula*	red	14.24 ±1.03	(SEM)	(n=35)	Lou et al.	2002	41
		white	28.92 ±0.84	(SEM)	(n=25)	Lou et al.	2002	41
Amphibians								
Salamander	*Ambystoma tigrinum n.*	leg extensor	33.9 ±2.8	(SEM)	(n=11)	Else et al.	1987	42
Tree frog	*Hyla chrysoscelis*	tensor chodarum	5.50 ±1.99	(95%)	(n=10)	McLister et al.	1995	43
		sartorius	25.22 ±3.82	(95%)	(n=6)	McLister et al.	1995	43
	Hyla cinera	tensor chodarum	18.05 ±11.78	(95%)	(n=3)	McLister et al.	1995	43
		sartorius	28.51 ±6.90	(95%)	(n=3)	McLister et al.	1995	43
	Hyla versicolor	tensor chodarum	9.44 ±1.67	(95%)	(n=16)	McLister et al.	1995	43
		sartorius	24.7 ±3.65	(95%)	(n=8)	McLister et al.	1995	43
	Osteopilus septenrionali	sartorius	24.1 ±2.3	(95%)	(n=9)	Peplowski et al	1997	44
Leopard frog	*Rana pipiens*	semimembranosus	25.5 ±1.12	(SEM)	(n=12)	Lutz et al.	1996	45
		sartorius	27.0 ±1.0	(SEM)	(n=7)	Renaud et al.	1984	46
Toad	*Bufo americanus*	white iliofibularis	26.0			Johnston et al.	1987	47
		sartorius	32.3 ±1.2	(SEM)	(n=7)	Renaud et al.	1984	46

Chapter 3 The maximum tetanic tension P_0.

Common Name	Species	Muscle	P_0 [N/cm^2]				Author	Year	Refs. No.
Amphibian									
Clawed frog	*Xenopus laeveus*	iliofibularis	26.0				La¨nnergren	1978	48
		iliofibularis	31.7				La¨nnergren	1978	48
		1n	39.6	±5.4	(SD)	(n=8)	La¨nnergren	1987	49
		1s	33.7	±3.8	(SD)	(n=10)	La¨nnergren	1987	49
		2f	31.2	±3.6	(SD)	(n=6)	La¨nnergren	1987	49
		2s	30.0	±4.9	(SD)	(n=6)	La¨nnergren	1987	49
		2	32.4	±5.0	(SD)	(n=10)	La¨nnergren et	1982	50
		1	37.2	±4.9	(SD)	(n=10)	La¨nnergren et	1982	50
			21.5	±1.2	(SEM)	(n=14)	Altringham et	1996	51
Toad	*Bufo woodhousei*	white iliofibularis	26.0				Johnston et al.	1987	52
Cane toad	*Bufo marinus*	white iliofibularis	26.0				Johnston et al.	1987	52
Toad	*Bufo alvarius*	depressor mandibulae	29.1	±0.4	(SEM)	(n=19)	Lappin et al.	2006	53
	Bufo alvarius	abductor indicus longus	30	±5	(SEM)	(n=5)	Clark et al.	2006	54
Australian Rocket Frog	*Litoria nasuta*	plantaris longus	39.0				James, Wilson	2008	55
Bullfrog	*Rana Catesbeina*	iliofibularis	29.2	±1.9	(SEM)	(n=5)	Hetherington	1987	56
		Opercularis	12.1	±1.4	(SEM)	(n=5)	Hetherington	1987	56
Edible frog	*Rana esculenta*	semitendinosus	33.4				Cecchi et al.	1978	57
	Rana esculenta	tibialis anterior	25.8	±0.3	(SEM)	(n=5)	Piazzesi et al.	2003	58
Reptiles									
Lizard	*Sceloporus occidentalis*	white iliofibularis	18.7	±8.4	(SEM)	(n=13)	Marsh et al.	1986	59
Desert iguana	*Dipsosaurus dorsalis*	white iliofibularis	21.4	±1.0	(SEM)	(n=25)	Marsh	1988	60
		white iliofibularis	33.0	±2.1		(n=16)	Johnston et al.	1984	22
		red iliofibularis	32.5	±1.5		(n=37)	Johnston et al.	1984	22
Terrapin	*Psuedomys scripta*	fast glycolytic	18.4	±0.5	(SEM)	(n=17)	Mutungi et al.	1987	11
		fast ox/glycolytic	12.0	±0.3	(SEM)	(n=16)	Mutungi et al.	1987	11
		slow	7.06	±0.03	(SEM)	(n=19)	Mutungi et al.	1987	11
Crocodile	*Crocodylus porosus*	caudofemoralis	23	±1	(SEM)	(n=8)	Seebacher et al	2008	61
Birds									
Quail	*Coturnix chinensis*	pectoralis	13.1	±0.54	(SEM)	(n=8)	Askew et al.	2001	62
Starling	*Sturnus vulgaris*	pectoralis	12.3				Biewener et al.	1992	63
Chicken	*Gallus domesticus*	white pectoralis	16.5				Reiser et al.	1996	64
		red pectoralis	17.4				Reiser et al.	1996	64
		latissimus dorsi	12.6				Reiser et al.	1996	64
Mammals									
Korean bat	*Murina leucogaster*	biceps brachii	15.74	±1.06	(SD)	(n=7)	Choi et al.	1998	65
Guinea pig	*Cavia porcellus*	soleus	14.7				Asmussen et al	1989	66
Rat	*Rattus norvegicus*	diaphragm	20.5				Johnson et al.	1994	67
		1	7.3				Schiaffino et a	1996	68
		2a	10.7				Schiaffino et a	1996	68
		2x	9.5				Schiaffino et a	1996	68
		2b	10.6				Schiaffino et a	1996	68
	(hybrid)	semimembranosus	17.1	±0.3	(SD)	(n=121)	Lowe et al.	2004	69
Wistar rat		extensor digitorum long	28.1	±2.5	(SD)	(n=5)	Close	1969	70
Rabbit	*Oryctolagus cuniculus*	inferior oblique	3.96	±0.37	(SEM)	(n=9)	Asmussen et al	1994	71
		1	10.6				Schiaffino et a	1996	68
		2a	11.6				Schiaffino et a	1996	68
		2b	13.2				Schiaffino et a	1996	68
Mouse	(Swiss strain)	soleus	12.31	±1.66	(SEM)	(n=6)	Lichtwark et al	2010	72
Cat		medial gastrocnemius(F	28.4				Burke et al.	1973	73
Monkey	*Macaca mulatta*	soleus	18.0				Fitts et al.	2000	74
		gastrocnemius	18.4				Fitts et al.	2000	74
Horse	*Equus caballus*	1	8.4				Rome et al.	1990	75
		2a	9.7				Rome et al.	1990	75
		2b	12.0				Rome et al.	1990	75
Human	*Homo sapiens*	1	21.0				Larson et al.	1993	76
		2a	20.0				Larson et al.	1993	76
		2b	19.0				Larson et al.	1993	76

Table 3.1 Continued

Chapter 3   The maximum tetanic tension P_0.

Common Name	Species	Muscle	P_0 [N/cm^2]			Author	Year	Refs. No.
Crustaceans								
Crab	*Cancer branneri*	claw closer	103.2 ±3.5	(SEM)		Taylor	2000	P32
	Cancer antennarius	claw closer	86.6 ±6.2	(SEM)		Taylor	2000	
	Menippe mercenaria	claw closer	220.0			Blundon	1988	78
	Carcinus maenas	flagellar	6.0			Stokes et al.	1994	79
	Callinectes sapidus	claw closer (crusher)	63.8 ±17.8	(SD)	(n=18)	Govind et al.	1985	80
Lobster	*Homarus americanus*	fast abdominal	8.38 ±1.49	(SEM)	(n=3)	Jahromi et al.	1969	81
		slow abdominal	45.11 ±7.56	(SEM)	(n=4)	Jahromi et al.	1969	81
		claw closer	39.0			Elner et al.	1981	82
Norway lobster	*Nephrops norvegicus*	abdominal flexor	10.5 ±3.9		(n=6)	Holmes et al.	1999	83
Mollusks								
Scallop	*Argopecten irradians*	adductor	21.4			Olson et al.	1993	84
Squid	*Alloteuthis subulata*	mantle	26.2 ±1.6	(SEM)	(n=6)	Milligan et al.	1997	85
Squid	*Loligo pealei*	transverse (of arm)	46.83 ±9.12	(SD)	(n=5)	Kier et al.	2002	86
		transverse (of tentacle)	13.09 ±5.56	(SD)	(n=12)	Kier et al.	2002	86
Cuttlefish	*Sepia officinalis*	mantle	22.6 ±1.9	(SEM)	(n=7)	Milligan et al.	1997	85
blue mussel	*Mytilus edulis*	anterior byssus retractor	55.0 ±4.0	(SD)	(n=173)	Gilbert	1978	87
fan mussel	*Pinna fragilis*	posterior adductor	58.8			Abbott et al.	1956	88

Table 3.1   *Continued*

Table 3.1 Published data of maximum tetanic tension P_0 appeared in the study of contractile property of muscle over various species. The (95%) represents 95 % confidence limits. The *n* in the parenthesis is number of samples. The data are also shown graphically in Fig.3.2 and Fig.3.3.

Chapter 3　The maximum tetanic tension P_o.

References

1　Weber, E., Wagner's Handworterbuch der physiologie (Braunschweig, Vieweg, 1846).

2　Josephson, R. K., Contraction dynamics of flight and stridulatory muscles of tettigoniid insects. J. Exp. Biol. 108, 77-96(1984).

3　Johnston, I. A. and Salamonski, J., Power output and forcevelocity relationship of red and white muscle fibres from the pacific blue marlin (Makaira nigricans). J. Exp. Biol. 111: 171-177(1984).

4　Johnston, I. A. and Harrison, P., Contractile and Metabolic characteristics of muscle fibres from antarctic fish. J. Exp. Biol. 116, 223-236(1985).

5　Marsh, R. L. and Bennett, A. F., Thermal dependence of contractile properties of skeletal muscle from the lizard Sceloporus occidentalis with comments of methods for fitting and comparing force-velocity curves. J. Exp. Biol.　126, 63-77(1986).

6　Else, P. L. and Bennet, A. F., The thermal dependence of locomotor performance and muscle contractile function in the salamander Ambystoma tigrinum nebulosum. J. Exp. Biol.　128, 219-233(1987).

7　Langfeld, K. S., Altringham, J. D., and Johnston, I. A., Temperature and the force-velocity relationship of live muscle fibres from the teleost Myoxocephalus scorpius. J. Exp. Biol. 144, 437-448(1989).

8　Malamud, J. G. and Josephson, R. K., Force-velocity relationships of a locust flight muscle at different times during a twitch contraction. J. Exp. Biol. 159, 65-87(1991).

9　Seebacher, F. and James, R. S., Plasticity of muscle function in a thermoregulating ectotherm (Crocodylus porosus): biomechanics and metabolism. Am. J. Physiol. Regul. Integr. Comp. Physiol. 294, R1024-R1032(2008).

10　Josephson, R. K., Malamud, J.G. and Stokes, D.R., Power output by an synchronous flight muscle from a beetle. J. Exp. Biol. 203: 2667-2689(2000).

11　Mutungi, G. and Johnston, I. A., The effects of temperature and pH on the contractile properties of skinned muscle fibres from the terrapin, Pseudemys scripta elegans. Journal of Experimental Biology 128: 87-105(1987).

12　Marsh, R. L., Ontogenesis of contractile properties of skeletal muscle and sprint performance in the lizard Dipsosaurus dorsalis., J. Exp. Biol. 137: 119-139(1988).

13　Lännergren, J., The force-velocity relation of isolated twitch and slow muscle fibres of Xenopus laevis. J. Physiol. 283,　501-521(1978).

14　Hutchinson, J. R. and Garcia, M., *Tyrannosaurus* was not a fast runner. Nature 415, 1018-1021(2002).

Chapter 3   The maximum tetanic tension P_o.

15 Johnston, I. A., Sustained force development specializations and variation among the vertebrates, J. Exp. Biol. 115, 239-251(1985).

16 Johnston, I. A. and Gleeson, T. T., Thermal dependence of contractile properties of red and white fibres isolated from the iliofibularis muscle of the desert iguana (Dipsosaurus dorsalis). J. Exp. Biol. 113, 123-132(1984).

17 Lännergren, J., Contractile properties and myosin isoenzymes of various kinds of Xenopus twitch muscle fibers. J. Musc. Res. Cell Motil. 8, 260--273(1987).

18 James, R. S. and Wilson, R. S., Explosive Jumping: Extreme Morphological and Physiological Specializations of Australian Rocket Frogs (Litoria nasuta). Physiol. Biochem. Zool. 81, 176-185(2008).

19 Hutchinson, J. R., Biomechanical modeling and sensitivity analyis of bipedal running ability. II. Extinct taxa, J. Morph. 262, 441-461(2004).

20 Gatesy, S. M., Baker, M., and Hutchinson, J. R., Constraint-Based Exclusion of Limb Poses for Reconstructing Theropod Dinosaur Locomotion. J. Vert. Paleo. 29, 535-544(2009).

21. Marden, J. H., Evolutionary adaptation of contractile performance in muscle of ectothermic winter-flying moths. J. Exp. Biol. 198, 2087–2094(1995).

22. Josephson, R. K., Contraction dynamics of flight and stridulatory muscles of tettigoniid insects. J. Exp. Biol. 108, 77–96(1984).

23. Fitzhugh, G. H. and Marden, J. H., Maturational changes in troponin T expression, Ca^{2+}–sensitivity and twitch contraction kinetics in dragonfly flight muscle. J. Exp. Biol. 200, 1473–1482(1997).

24. Marden, J. H., Fitzhugh, G. H., Girgenrath, M., Wolf, M. R., and Girgenrath, S., Alternative splicing, muscle contraction and intraspecific variation: associations between troponin T ranscripts, Ca2+ sensitivity and the force and power output of dragonfly flight muscles during oscillatory contraction. J. Exp. Biol. 204, 3457-3470(2001).

25. Josephson, R. K. and Ellington, C. P., Power output from a flight muscle of the bumblebee Bombus terrestris. I. Some features of the dorso-ventral flight muscle. J. Exp. Biol. 200, 1215–1226(1997).

26. Malamud, J. G. and Josephson, R. K., Force-velocity relationships of a locust flight muscle at different times during a twitch contraction. *J. Exp. Biol.* 159, 65–87(1991).

27. Josephson, R. K., Malamud, J.G., and Stokes, D.R., Power output by an synchronous flight muscle from a beetle. J. Exp. Biol. 203: 2667-2689(2000).

28. Marden, J. H. Evolutionary adaptation of contractile performance in muscle of ectothermic winter-flying moths. J. Exp. Biol. 198: 2087–2094(1995).

29. Full, R. J., Stokes, D. R., Ahn, A. N., and Josephson, R. K., Energy absorption during running by leg muscles in a cockroach. J. Exp. Biol. 201, 997–1012(1998).

Chapter 3 The maximum tetanic tension P_o.

30. Rome, L. C. and Swank, D., The influence of temperature on power output and scup red muscle during cyclical length changes. J. Exp. Biol. 171, 261-281(1992).

31. Josephson, R. K., Contraction dynamics of flight and stridulatory muscles of tettigoniid insects. J. Exp. Biol. 108, 77–96(1984).

32. James, R. S., Cole, N. J., Davies, M. L. F., and Johnston, I. A., Scaling of intrinsic contractile properties and myofibrillar protein composition of fast muscles in the fish Myoxocephalus scorpius L. J. Exp. Biol. 201, 901–912(1998).

33. Franklin, C. E. and Johnston, I. A., Muscle power output during escape responses in an Antarctic fish. J. Exp. Biol. 200, 703-712(1997).

34. Johnston, I. A. and Brill, R., Thermal dependence of contractile properties of single skinned muscle fibres from Antarctic and various warm water marine fishes including skipjack tuna (Katsuwonus pelamis) and kawakawa (Euthynnus affinis). J. Comp. Physiol. B 155, 63–70(1984).

35. Rome, L. C., Syme, D. A., Hollingworth, S., Lindstedt, S. L., and Baylor, S. M., The whistle and the rattle: the design of sound producing muscles. Proc. Natl. Acad. Sci. USA 93, 8095–8100(1996).

36. Coughlin, D. J., Zhang, G., and Rome, L. C., Contraction dynamics and power production of pink muscle of the scup (Stenotumuschrysops). J. Exp. Biol. 199, 2703–2712(1996).

37. Johnston, I. A. and Salamonski, J., Power output and forcevelocity relationship of red and white muscle fibres from the pacific blue marlin (Makaira nigricans). J. Exp. Biol. 111: 171–177(1984).

38. Johnston, I. A. and Harrison, P., Contractile and Metabolic characteristics of muscle fibres from antarctic fish. J. Exp. Biol. 116, 223-236(1985).

39. Curtin, N. A. and Woledge, R. R., Power output and force-velocity relationship of live fibres from white myotomal muscle of the dogfish, Scyliorhinus Canicula. J. Exp. Biol. 140, 187-197(1988).

40. Langfeld, K. S., Altringham, J. D. and Johnston, I. A., Temperature and the force-velocity relationship of live muscle fibres from the teleost Myoxocephalus scorpius. J. Exp. Biol. 144, 437-448(1989).

41. Lou, F., Curtin, N. A., and Woledge, R. C., Isometric and isovelocity contractile performance of red muscle fibres from the dogfish Scyliorhinus canicula. J. Exp. Biol. 205, 1585-1595(2002).

42. Else, P. L. and Bennet, A. F., The thermal dependence of locomotor performance and muscle contractile function in the salamander Ambystoma tigrinum nebulosum. J. Exp. Biol. 128, 219–233(1987).

43. McLister, J. D, Stevens, E. D., and Bogart, J. P., Comparative contractile dynamics of calling and locomotor muscles in three hylid frogs. J. Exp. Biol. 198, 1527–1538(1995).

Chapter 3 The maximum tetanic tension P_o.

44. Peplowski, M. M. and Marsh, R. L., Work and power output in the hindlimb muscles of cuban tree frogs Oseopilus septentrionalis during jumping. J. Exp. Biol. 200, 2861–2870(1997).
45. Lutz, G. J. and Rome, L. C., Muscle function during jumping in frogs. II. Mechanical properties of muscle: implications for system design. Am. J. Physiol. - Cell Physiol. 271, C571–C578(1996).
46. Renaud, J. M. andStevens, E. D., The extent of short-term and long-term compensation to temperature shown by frog and toad sartorius muscle. J. Exp. Biol. 108: 57-75(1984).
47. Johnston, I. A. and Gleeson, T. T., Thermal dependence of contractile properties of red and white fibres isolated from the iliofibularis muscle of the desert iguana (Dipsosaurus dorsalis). J. Exp. Biol. 113, 123-132(1984).
48. Lännergren, J., The force-velocity relation of isolated twitch and slow muscle fibres of Xenopus laevis. J. Physiol. 283, 501–521(1978).
49. Lännergren, J., Contractile properties and myosin isoenzymes of various kinds of Xenopus twitch muscle fibers. J. Musc. Res. Cell Motil. 8, 260–273(1987).
50. Lännergren, J., Lindblom, P., and Johansson, B., Contractile properties of two varieties of twitch muscle fibtres in Xenopus laevis. Acta Physiol. Scand. 114, 523-535(1982).
51. Altringham, J. D. , Morris, T. James, R. S., and Smith, C. I., Scaling effects on muscle function in fast ans slow muscles of Xenopus laevis. Exp. Biol. Online 1, 6-12 (1996).
52. Johnston I. A. and Gleeson T. T., Effects of temperature on contractile properties of skinned muscle fibers from three toad species. Am. J. Physiol. Regul. Integr. Comp. Physiol. 252, R371–R375(1987)
53. Lappin, A. K., Monroy, J. A., Pilarski, J. Q., Zepnewski, E. D., Pierotti, D. J., and Nishikawa, K. C., Storage and recovery of elastic potential energy powers ballistic prey capture in toads. J. Exp. Biol. 209, 2535-2553(2006).
54. Clark, D. L. and Peters, S. E., Isometric contractile properties of sexually dimorphic forelimb muscles in the marine toad Bufo Marinus Linnaeus 1758: functional analysis and implications for amplexus. J. Exp. Biol. 209: 3448-3456(2006).
55. James, R. S. and Wilson, R. S., Explosive Jumping: Extreme Morphological and Physiological Specializations of Australian Rocket Frogs (Litoria nasuta). Physiol. Biochem. Zool. 81, 176-185(2008).
56. Hetherington, T. E., Physiological features of the opercularis muscle and their effects onvibration sensitivity in the bullfrog Rana catesbeiana. J. Exp. Biol. 131: 189-204(1987).
57. Cecchi, G., Colomo, F. and Lombardi, V., Force-velocity relation in normal and nitrate-treated frog single muscle fibres during rise of tension in an isometric tetanus. J. Physiol. 285, 257-273(1978).

Chapter 3 The maximum tetanic tension P_0.

58. Piazzesi, G., Reconditi, M., Koubassova, N., Decostre, V., Linari, M., Lucci, L., and Lombardi, V., Temperature dependence of the force-generating process in single fibres from frog skeletal muscle. J. Physiol. 549, 93-106(2003).
59. Marsh, R. L. and Bennett, A. F., Thermal dependence of contractile properties of skeletal muscle from the lizard Sceloporus occidentalis with comments of methods for fitting and comparing force-velocity curves. J. Exp. Biol. 126, 63–77(1986).
60. Marsh, R. L., Ontogenesis of contractile properties of skeletal muscle and sprint performance in the lizard Dipsosaurus dorsalis. J. Exp. Biol. 137, 119–139(1988).
61. Seebacher, F. and James, R. S., Plasticity of muscle function in a thermoregulating ectotherm (Crocodylus porosus): biomechanics and metabolism. Am. J. Physiol. Regul. Integr. Comp. Physiol. 294, R1024-R1032(2008).
62. Askew, G. N. and Marsh, R. L., The mechanical power output of the pectoralis muscle of the blue-breasted quail (Coturnix chinensis): the in vivo length cycle and its implications for muscle performance. J. Exp. Biol. 204: 3587–3600(2001).
63. Biewener, A. A., Dial, K. P., and Goslow, G. E., Pectoralis muscle force and power output during flight in the starling. J. Exp. Biol. 164: 1–18(1992).
64. Reiser, P. J., Greaser, M. L., and Moss, R. L., Contractile properties and protein isoforms of single fibers from the chicken pectoralis red strip muscle. J. Physiol. 493, 553–562(1996).
65. Choi, I., Cho, Y., Oh, Y. K., Jung, N., and Shin, H., Behavior and muscle performance in heterothermic bats. Physiol. Zool. 71, 257–266(1998).
66. Asmussen, G. and Maréchal, G., Maximal shortening velocities, isomyosins and fibre types in soleus muscle of mice, rats, and guinea-pigs. J. Physiol. 416, 245–254(1989).
67. Johnson, B. D., Wilson, L. E., Zhan, W. Z., Watchko, J. F., Daood, M. J., and Sieck, G. C., Contractile properties of the developing diaphragm correlate with myosin heavy chain phenotype. J. Appl. Physiol. 77, 481–487(1994).
68. Schiaffino, S. and Reggiani, C., Molecular diversity of myofibrillar proteins: gene regulation and functional significance. Physiol. Rev. 76, 371–423(1996).
69. Lowe, D. A., Warren, G. L., Snow, L. M., Thompson, L. V., and Thomas, D. D., Muscle activity and aging affect myosin structural distribution and force generation in rat fibers, J. Appl. physiol. 96, 498-506(2004).
70. Close, R. I., Dynamic properties of fast and slow skeletal muscles of the rat after nerve cross-union, J. Physiol. 204: 331-346(1969).
71. Asmussen, G., Beckers-Bleukx, G., and Maréchal G., The force-velocity relation of the rabbit inferior oblique muscle: influence of temperature. Pflüg. Arch. 426, 542–547(1994).
72. Lichtwark, G. A. and Barclay, C. J., The influence of tendon compliance on muscle power output and efficiency during cyclic contractions. J. Exp. Biol. 213, 707-714(2010).

Chapter 3   The maximum tetanic tension P_o.

73. Burke, R. E. and Tsairis, P., Anatomy and innervation ratios in motor units of cat gastrocnemius, J. Physiol. 234, 749-765(1973).
74. Fitts, R. H., Desplanches, D., Romatowski, J. G., and Widrick, J. J., Spaceflight effects on single skeletal muscle fiber function in the rhesus monkey. Am. J. Physiol. Regul. Integr. Comp. Physiol. 279, R1546–R1557(2000).
75. Rome, L.C., Sosnicki, A. A., and Goble, D. O., Maximum velocity of shortening of three fibre types from horse soleus muscle: implications for scaling with body size. J. Physiol. 331, 173-185(1990).
76. Larson, L. and Moss, R. L., Maximum velocity of shortening in relation to myosin isoform composition in single fibres from human skeletal muscles. J. Physiol. 472, 595-614(1993).
77. Taylor, G. M., Maximum force production: why are crabs so strong? Proc. R. Soc. Lond. B Biol. Sci. 267, 1475-1480(2000).
78. Blundon, J. A., Morphology and muscle stress of chelae of temperate and tropical stone crabs Menippe mercenaria. J. Zool. 215, 663–673(1988).
79. Stokes, D. R. and Josephson, R. K., Contractile properties of a high-frequency muscle from a crustacean. II. Contraction kinetics. J. Exp. Biol. 187, 275–293(1994).
80. Govind, C. K. and Blundon, J. A., Form and function of the asymmetric chelae in blue crabs with normal and reversed handedness. Biol. Bull. 168, 321–331(1985).
81. Jahromi, S. S. and Atwood, H. L., Correlation of structure, speed of contraction, and total tension in fast and slow abdominal muscle fibers of the lobster. J. Exp. Zool. 171, 25–38(1969).
82. Elner, R. W. and Campbell, A., Force, function and mechanical advantage in the chelae of the American lobster Homarus americanus (Decapoda: Crustacea). J. Zool. 193, 269–286(1981).
83. Holmes, J. M., Hilber, K., Galler, S., and Neil, D. M., Shortening properties of two biochemically defined muscle fiber types of the Norway lobster Nephrops norvegicus L. J. Musc. Res. Cell Motil. 20, 265–278(1999).
84. Olson, J. M. and Marsh, R. L., Contractile properties of the striated adductor muscle in the bay scallop Argopecten irradians at several temperatures. J. Exp. Biol. 176, 175–193(1993).
85. Milligan, B,, Curtin, N. A., and Bone, Q., Contractile properties of obliquely striated muscle from the mantle of squid (Alloteuthis subulata) and cuttlefish (Sepia officinalis). J. Exp. Biol. 200: 2425–2436(1997).
86. Kier, W. M. and Curtin, N. A., Fast muscle in squid (Loligo pealei): contractile properties of a specialized muscle fibre type. J. Exp. Biol. 205: 1907-1916(2002).
87. Gilbert, S. H. Tension and heat production during isometric contractions and shortening in the anterior byssus retractor muscle of Mytilus edulis., J. Physiol. 282: 7-20(1978).
88. Abbott, B. C. and Lowy, J., Mechanical properties of Pinna adductor muscle. J. Mar. Biol. Assoc. U. K. 35, 521-530(1956).

Chapter 3   The maximum tetanic tension P_0.

89. Bates, K. T., Manning, P. L., Hodgetts, D., and Sellers, W. I., Estimating Mass Properties of Dinosaurs Using Laser Imaging and 3D Computer Modelling, PLoS ONE 4 (2): e4532 doi:10.1371/journal.pone.0004532(2009).
90. Hutchinson, J. R., Anderson, F. C., Blemker, S. S., and Delp, S. L., Analysis of hindlimb muscle moment arms in Tyrannosaurus rex using a three-dimensional musculoskeletal computer model: implications for stance, gait, and speed, Paleobiology 32: 676-701(2005).
91. Sellers, W. I. and Manning, P. L., Estimating dinosaur maximum running speeds using evolutionary robotics, Proc. Roy. Soc. B 274: 2711-2716(2007).

Chapter 4 Maximum muscle stress and specific tension σ

In the previous chapter isometric tetanic tension P_o was introduced. For determining the value several bundles of fibres are dissected from animal, and shortening force with constant length is measured. Cross-sectional area of dissected bundles of fibres is small as the range of $1 \times 10^{-2} \sim 1 \times 10^{-5}$ cm².

On the other hand another quantity called maximum muscle stress σ has been measured since 1846, which was mentioned in Chap.2. In this chapter this quantity i.e., maximum muscle stress σ is introduced in detail. At present, sports science is a major field that this quantity is applied for analysis of physical motion. For measuring σ, modern techniques such as MRI and real-time ultrasonic measurement have been used. It seems that the use of such modern technique may lead us to a bridge over the knowledge of muscle fibres level and individual movement level. However, it is said that there is still a great gap of understanding on muscle strength between these two levels.

4.1 Maximum muscle stress σ

In 1846, Weber proposed that absolute muscle force should be regarded as tension per cross section of muscle, and not on its length (Weber, 1846). After that, many researchers agreed with it, however, the measurement of its absolute value has been turned out as a hard task. Difficulties in the measurement of physiological cross section, lever arm length and the other factors resulted in a wide range of absolute value as 11~180 N/cm². In the early date σ was reported as,

61 N/cm²	Plantar flexors	Herman	1898
51.5 N/cm²	Trunk	Reys	1915
36-41 N/cm²	Plantar flexors	Haxton	1944
76.8~90.3 N/cm²	Knee flexors/extensors	Morris	1948
90~153 N/cm²	Elbow flexors/extensors	Morris	1948

In the literature, σ value was also reported as 98 N/cm² (Johnson, 1903), 59-98 N/cm² (Fick, 1904), 35 N/cm² (Rechlinghausen, 1920), 35 N/cm² (Arkin, 1938).

From 1960's measurements of physiological data such as cross-sectional area and lever arm length have been carried out by modern techniques such as X-ray and ultrasonic. In the next sub-section, typical method of measuring the maximum muscle force σ is introduced.

Chaper 4 Maximum muscle stress and specific tension σ

4.2 Measuring the Maximum Muscle Force σ N/cm²

In 1968, Ikai and Fukunaga published experimental observation data of σ N/cm² for elbow flexors. X-Ray photograph was used to determine lever arm ratio, and ultrasonic photograph was used to determine cross-sectional area.

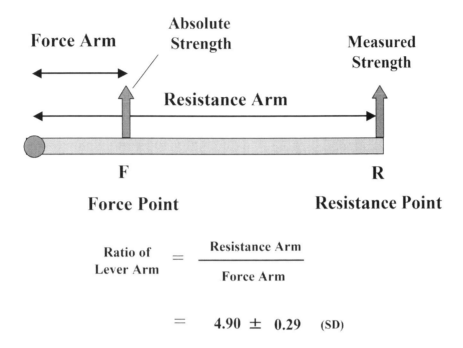

Fig.4.1 X-ray photograph of the arm flexor at the right angle of the elbow joint in sitting position (Ikai and Fukunaga, 1968). Each position is denoted as O; Joint, F; Force point, R; Resistance Point. The muscle force must be calculated at force point F. Actual measurement of arm strength was observed at resistance point R. Then, muscle force was calculated by a product of arm strength and ratio of lever arm. The ratio of lever arm was defined as resistance arm/force arm, which was 4.90±0.29 (SD) in this case.

For a subject, a cloth belt is attached at the wrist which is shown as R in Fig. 4.1. The subject contracted the muscle against the cloth belt with maximum effort. Any especially procedure was not used to get the maximum strength in this experiment.

From the picture of X-ray, the lever arm ratio is calculated as the ratio of resistance arm OR divided by force arm OF, i.e., lever arm ratio=resistance arm/force arm. Then, force of upper arm extensor F was determined as F multiplied by the lever arm ratio. The value of the lever arm ratio was obtained as 4.90±0.29 (SD) in this case.

Chaper 4 Maximum muscle stress and specific tension σ

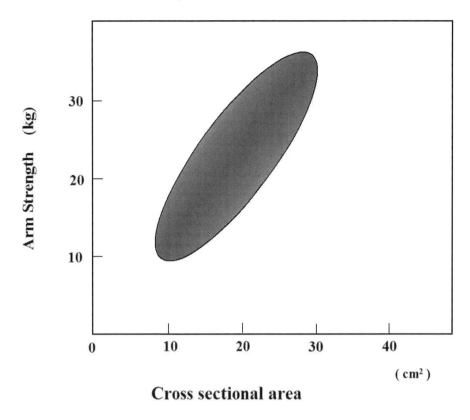

Fig.4.2   The relationship between arm strength in the unit of kg and cross-sectional area of flexor of the upper arm (Ikai and Fukunaga, 1968). The arm strength kg multiplied by gravitational constant 9.8 m/s^2 yields arm force in the unit of N. Although the data are scattered, the linear relation of arm strength to cross sectional area is observed, For ordinary 20 years old 12 subjects, mean and standard deviation of muscle strength per cross sectional area are determined as 6.7 $\pm$ 1.1 kg/cm^2, which leads to the maximum muscle stress of elbow flexor as σ =66$\pm$11 N/cm^2

Ultrasonic photography was used to obtain cross-sectional area of muscle. Under consideration of the structure of subcutaneous fat and fascia, muscle of the upper arm was divided into flexors (biceps brachii and brachialis) and extensors. Fig.4.2 shows a relation between cross-sectional area of flexors and arm strength. A product of arm strength (kg) and gravitational constant 9.8 m/s² divided by cross-sectional area yields maximum muscle stress σ N/cm². Typical value of σ was reported as,

$$\sigma = 66 \pm 11 \text{ N/cm}^2 \quad \text{(for 20 yrs male subjects (}N\text{=12))} \quad .$$

As is observed from Fig.4.2, the values of σ are distributed in a wide range. The data were collected for young/old, male/female and ordinary/Judomen.

Chaper 4 Maximum muscle stress and specific tension σ

Ikai and Fukunaga reported that σ values were distributed in a range as,

$$\sigma = 40 \sim 80 \quad N/cm^2 \quad .$$

4.3 Using MRI scanning for determining specific tension σ N/cm²

From 1990's modern scanning techniques have been developed, and those enabled capturing images of muscle in situ. In this subsection Kawakami et al.'s work is introduced that achieved detailed scanning of upper extremities, and determined the constant of force per cross-sectional area. From 1980's this constant has been called specific tension instead of maximum muscle stress. The reason is that the constant has been recognized to be muscle specific one instead of universal constant. This situation is described in the next subsection in detail.

 Note; From 1980's, σ has been called **specific tension** instead of maximum muscle stress. Because, it is not universal constant, but, muscle specified quantity. This difference is discussed in the subsection 4.4.

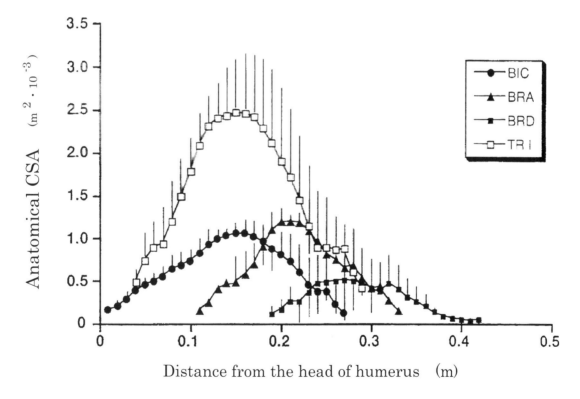

Fig.4.3   Series anatomical cross-sectional area (ACSA) along the length of the upper extremity, (Kawakami et al., 1994). The horizontal axis shows the distance from the head of humerus. The right end corresponds to the wrist joint. Elbow joint is located around 27 to 30cm from 0 point (head of humerous). The abbreviations are BIC (Biceps brachii), BRA (Brachialis), BRD (Brachioradialis) and TRI (Triceps brachii).

Chaper 4 Maximum muscle stress and specific tension σ

Figure 4.3 shows series anatomical cross-sectional areas (ACSA) along the length of the upper extremity using MRI (Magnetic Resonance Imaging), (Kawakami et al., 1994) . The horizontal axis shows the distance from the head of the humerus. The right end is the wrist joint. Elbow joint is located around 27 to 30 cm from 0 point (head of humerous).

Obviously, the shape of muscle is no longer measured as rectangular, but recognized as flexible object varying ACSA (anatomical cross-sectional area) along the coordinate (distance form the origin). By summing up of ACSA, we know muscle volume V by multiplying interspaced distance.   Specific tension is a quantity defined by force divided by physiological cross-sectional area (PCSA). Then, each quantity is calculated by its definition based on the measured data. Kawakami et al. used the definition of PCSA as follows,

$$\text{For elbow flexor muscles} \quad \text{PCSA} = \frac{V}{l_f} \quad , \tag{4.1}$$

$$\text{For elbow extensor muscles} \quad \text{PCSA} = \frac{V}{l_f}\cos\theta \quad , \tag{4.2}$$

where l_f and θ are fibre length and pennation angle, respectively. The fibre length l_f is calculated by,

$$l_f = l_{mu} \cdot k_m \quad , \tag{4.3}$$

where l_{mu} is the distance between the most proximal and the most distal images in which the muscle is visible. The coefficient k_m is derived from the previous study by Edgerton et al. (Edgerton et al., 1986) as,

$$k_m = \frac{l_f(cavaders)}{l_{mu}(cavaders)} . \tag{4.4}$$

The value of pennation angle θ was also employed from the previous study by Amis (Amis, 1979).

Maximal voluntary isometric, concentric, and eccentric strength of elbow flexors and extensors were measured using an isokinetic dynamometer. The device had a motor-driven lever arm that rotated around an axis at a constant velocity.

Chapter 4 Maximum muscle stress and specific tension σ

The measurements of elbow flexion and extension torque were carried out with the subjects seated, with their dominant arm supported on the horizontal plane on a padded table. Isometric torque was measured at elbow joint angles from 0.873 to 1.920 (rad). Concentric and eccentric torque were defined as the torque at which the maximal isometric torque was observed (flexion 1.571 rad, extension 1.222 rad).

Tendon force F_t was defined as,

$$F_t = \frac{\text{torque}}{\text{moment arm}} \qquad (4.5)$$

From F_t and PCSA (physiological cross-sectional area), specific tension σ N/cm² is calculated as,

$$\sigma \text{ (Specific tension)} = \frac{F_t}{\text{PCSA}} \qquad (4.6)$$

The mean value and standard deviation of each quantity for five subjects were as follows;

	Flexors	Extensors
Torque [N·m]	59.9±14.5	67.7±18.1
Moment arm [cm]	3.4±0.2	2.2±0.1
Pennation angle [rad]	0	0.215
PCSA [m²·10⁻³]	2.5±0.44	4.68±1.05

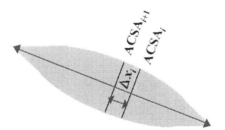

$$V = \sum_{i=1}^{N} \text{ACSA}_i \cdot \Delta x_i$$

Fig.4.4 The summation of the product of anatomical cross-sectional area (ACSA) and slice distance Δx yields muscle volume V.

Chapter 4 Maximum muscle stress and specific tension σ

The flexors include biceps brachii, brachialis, brachioradialis. The extensor is triceps brachii. Moment arm is the distance from the rotation centre to the line through the middle of the muscle and tendon. The relations of each quantity are graphically shown in Fig. 4.5, Fig. 4.6 and Fig. 4.7.

By using MRI, structure of the muscle is captured as flexible object varying cross-sectional area along the axis, not cylinder. Then, a summation of the product of anatomical cross-sectional area (ACSA) and slice distance Δx yields muscle volume V as shown in Fig.4.4.

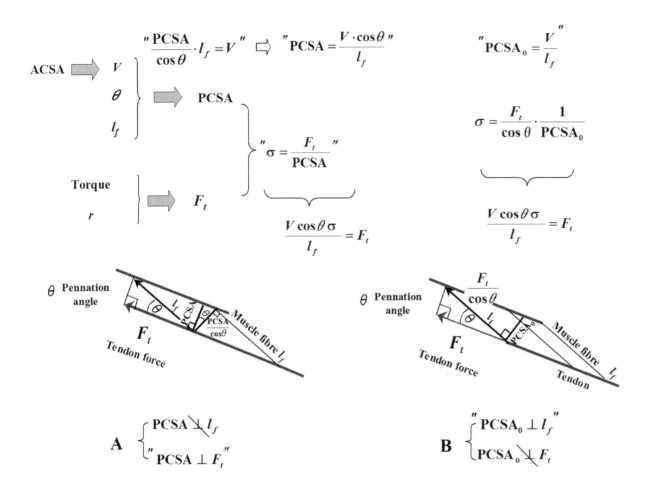

Fig.4.5 Two different interpretation of PCSA (physiological cross-sectional area) for pennetrated muscle. For the case of **A**, PCSA is perpendicular to the direction of tendon force F_t. For the case of **B**, PCSA$_0$ is perpendicular to the direction of muscle fibre l_f. The quotation " " expresses definition. The detailed explanation is given in the text.

Two different interpretations of physiological cross-sectional area (PCSA) are possible as shown in Fig.4.5.

Chapter 4 Maximum muscle stress and specific tension σ

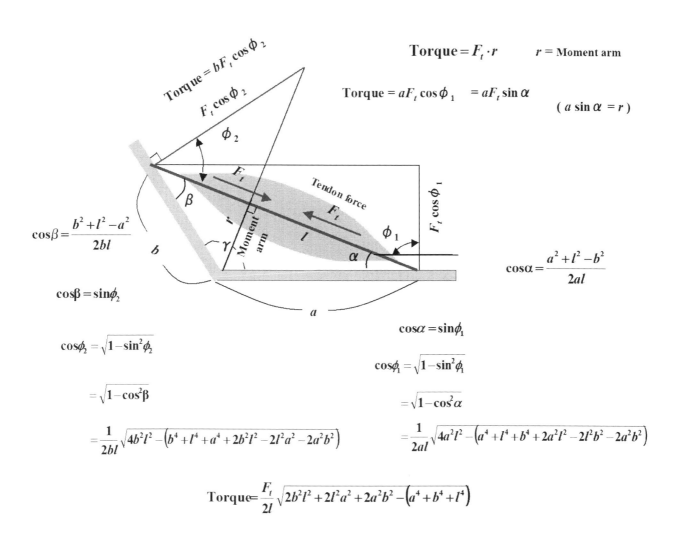

Fig.4.6 Geometrical relation of each length and the force. Torque = Force · Moment arm = $F_t r = a F_t \cos\phi_1 = b F_t \cos\phi_2$

Definition A

This is the one described in this section. In this case, the plane of PCSA is perpendicular to the direction of F_t and also tendon, but not to the direction of muscle fibre l_f. Then, σ is given by $\sigma = \dfrac{F_t}{\text{PCSA}}$. The plane of $\dfrac{\text{PCSA}}{\cos\theta}$ is perpendicular to the direction of l_f, which yields $V = \dfrac{\text{PCSA}}{\cos\theta} \cdot l_f$. Thus, $\text{PCSA} = \dfrac{V \cdot \cos\theta}{l_f}$ is obtained.

Chaper 4 Maximum muscle stress and specific tension σ

Definition B

For this case, PCSA₀ is defined by the volume V divided by the fibre length l_f, $PCSA_0 = \dfrac{V}{l_f}$. The PCSA₀ plane is perpendicular to the direction of muscle fibre l_f. The force acting to the muscle fibres is $\dfrac{F_t}{\cos\theta}$.

Then, σ is defined by $\sigma = \dfrac{F_t}{\cos\theta} \cdot \dfrac{1}{PCSA_0}$.

In both cases the final relation becomes the same as $\dfrac{V\cos\theta}{l_f} = F_t$.

Fig.4.6 displays the relation of each length. Torque of joint angle is defined by the product of force F_t and moment arm r. In Fig. 4.6, $a \cdot \sin\alpha$ is r. And, $\sin\alpha$ is <u>equal</u> to $\cos\phi_1$ when $\alpha + \phi_1 = \dfrac{\pi}{2}$. Then, torque becomes $aF_t\cos\phi_1$. This is identical to $bF_t\cos\phi_2$, which is relevant as shown in Fig.4.6,

$$\text{Torque} = \text{Force} \cdot \text{Moment arm} = F_t r = aF_t\cos\phi_1 = bF_t\cos\phi_2 \quad . \quad (4.7)$$

Figure 4.7 shows the relationships between the velocity (muscle fibre velocity) and the specific tension (maximum muscle stress σ), (Kawakami, et. al., 1994). In this experiment concentric and eccentric force were measured, then, velocity-dependent specific tension was obtained. Positive velocities show concentric actions, and negative velocities show eccentric actions. The velocity of zero represents isometric condition. Thus, the specific tension at $v = 0$ represents σ [N/cm²] which is defined as maximum muscle stress.

The specific tension at $v=0$ is $\sigma = 600 \sim 750$ kN/m² which corresponds to $60 \sim 75$ N/cm². This result in general coincides with the previous result obtained by the result (Ikai and Fukunaga, 1968).

In 1995, Buchanan reported comparatively similar or higher value of σ for elbow flexors and extensors (Buchanan, 1995). Typical values in the report are as the followings;

	Flexors	Extensors
Torque [N·m]	60	40
PCSA [cm²]	21.7	32.4
Specific tension σ [N/cm²]	99~148	43~91

Chaper 4 Maximum muscle stress and specific tension σ

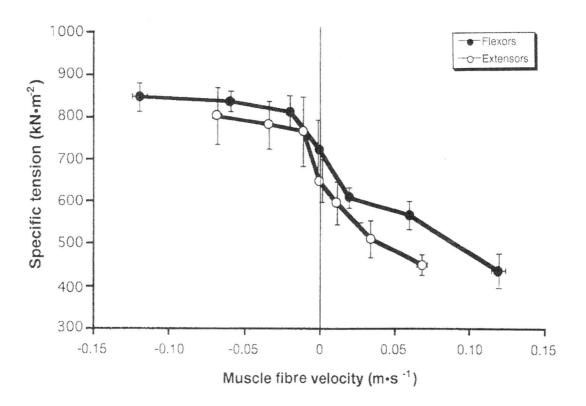

Fig.4.7 Relation between muscle fibre velocity and specific tension (Kawakami, etl.al. 1994). The velocity of zero represents isometric condition. Positive and negative velocities show concentric and eccentric actions, respectively.

He experimentally measured elbow torque, and obtained σ together with published results of anatomical parameters. He stated that the specific tension is not a constant value, but joint angle dependent. He also warned of the usage of specific tension σ as follows;

"These results indicate that biomechanical models using this assumption could be introducing errors of up to 50%. This technique is often used to estimate muscle forces, and this work clearly shows that its application should be reconsidered in situations where accurate solutions are required."

Actually, the data of force v.s. cross-sectional area is very much scattered in Fig. 4.2. Many researchers have warned for the usage of σ. For example Maughan et al. who measured σ knee-extensor muscles in 1983 (Maughan et al., 1983). They reported value of σ as 9.49 ± 1.34 (SD) N/cm² for knee-extensor muscles in 25 male subjects. This value is considerably smaller than that of Ikai and Fukunaga, Kawakami et al, and Buchanan's. Maughan et al. stated as follows in the article;

Chapter 4 Maximum muscle stress and specific tension σ

"A wide variation in the ratio of strength to muscle cross-sectional are was observed. This variability may be a result of anatomical differences between subjects or may result from differences in the proportions of different fibre types in the muscles. The variation between subjects is such that strength is not useful predictive index of muscle cross-sectional area." (Maughan et al., 1983.)

As we will see in the next subsection the range of reported σ spans 9~180 N/cm². Why was such large variance on measuring σ brought out ? One possible answer for this question may lay in the difference of muscle groups for measuring. Ikai and Fukunaga, Kawakami et al., and Buchanan's work were accomplished for elbow. On the contrary, Maughan et al.'s work was accomplished for knee. In the next subsection, reported value of σ is summarized for each muscle group.

4.4 Why is the value of σ so scattered ?

In this subsection, the question why large variance of the value of σ has been reported is discussed. Table 4.1~4.4 are the data for human mandible and trunk, elbow, knee and foot, respectively. Table 4.5 summarizes the value of σ employed in simulation studies.

Author		σ (N/cm²)	PCSA (cm²)	F (N)	Year	Index
		σ for Trunk and Mandible				
Prium et al. (1980)	Human mandible	140±33 (90–180)	--	--	1980	a1
Reys	Human trunk	51.5	--	--	1915	b1
Shultz et al.	Human trunk	44	[59.3]	2610	1982	b2
Reid and Costigan	Human trunk	47.8±12.5	54.38±8.37	[2600]	1987	b3

Table 4.1 Observed value of maximum muscle stress σ for human mandible and trunk. The value with parenthesis shows range that appeared in the article. The value with square blacket is calculated value from the other quantity. For example, PCSA [59.3] cm² is calculated from σ =44 N/cm² and F=2610 N. The blank column is the one that is not explicitly shown in the article. The data are also shown graphically in Fig.4.8 with index which is shown at the right column.

Chaper 4 Maximum muscle stress and specific tension σ

σ for Elbow							
Author	Elbow flexors/ extensors	Muscles	σ (N/cm^2)	PCSA (cm^2)	F (N)	Year	Index
Morris	Elbow flexors	BIC, BRA BRD	90	22	1980	1948	c1
	Elbow extensors	Triceps brachii	153	23	3524		
Ikai and Fukunaga	Elbow flexors	BIC, BRA	66.7 (60-80)	20	1320	1968	c2
Hatze	Elbow extensors	Triceps brachii	87.1 "male" 69.7 "female"	---	---	1981	c3
Kawakami	Elbow flexors	BIC, BRA BRD	72	25	1800	1994	c4
	Elbow extensors	Triceps brachii	64	46.8	3000		
Buchanan	Elbow flexors	BIC,BRA BRD, PT	99-148	21.7 "average"	[2148-3212]	1995	c5
	Elbow extensors	ANC,TRI	43-91	32.4 "average"	[1993-2948]		
Koo et al.	Elbow flexors	LHB, SHB BRA,BRD	129.37 (98.29-167.58)	---	---	2002	c6
Langenderfer et al.	Elbow flexors	BIC, BRA BRD	141±20 (SD)	---	---	2005	c7

Table 4.2 Observed value of σ for elbow flexors and extensors. Abbreviations are; biceps brachii (BIC), brachialis (BRA), brachioradialis (BRD), triceps bracii (TRI), long and short head brachii (LHB, SHB), pronator teres (PT), anconeus (ANC).

σ for Knee							
Author	Elbow flexors/ extensors	Muscles	σ (N/cm^2)	PCSA (cm^2)	F (N)	Year	Index
Morris	Knee flexors	Biceps femoris+	90.3	----	----	1948	d1
	Knee extensors	Quadriceps femoris	76.8	----	----		
Maughan et al.	Knee extensors	Quadriceps femoris	9.49±1.34 "Male" 8.92±1.11 "Female"	83.2±12.3 "M" (59.7-106.4)	783±118 "M" (543-1024)	1983	d2
Narici et al.	Knee extensors	Quadriceps femoris	25.0±1.9	280.1	1452~1997	1992	d3
Akima et al.	Knee extensors		23.5	291.7	6851	1994	d4
	Knee flexors		33.4	110.6	3698		
Reeves et al.	Knee extensors	Vastus lateralis	32.1±7.4	29.1±8.4	939.3±347.8	2004	d5
O'Brien et al.	Knee extensors	Quadriceps femoris	55.02±11 "Men" 59.77±15.3 "Girls"	231.8±55.5	[12753]	2009	d6

Table 4.3 The observed value of σ for knee flexors and extensors.

Chapter 4 Maximum muscle stress and specific tension σ

σ for Plantar flexors and Dorsiflexors						
Author	Elbow flexors/ extensors	σ (N/cm^2)	PCSA (cm^2)	F (N)	Year	Index
Hermann	Plantar flexors	61	114.7	7020	1898	e1
Haxton	Plantar flexors	38±1	113	4300	1944	e2
Vandervoort and McComas	Plantar flexors	14	248	3563	1986	e4
Davies et al.	Plantar flexors	19.5 "old" 32.9 "young"	59	1900	1986	e5
Fukunaga et al.	Plantar flexors	11	337	3623	1996	e6
	Dorsiflexorss	24	35	832		
Wickiewicz	Plantar flexors	30	92	2769	1984	e7
	Dorsiflexors	47	17	792		

Table 4.4 The observed value of σ for plantar flexors and dorsiflexors

σ employed in simulation study					
Author	Elbow flexors/ extensors	Muscles	σ (N/cm^2)	Year	Index
McGill and Norman	Trunk extensors	Erector spinae	30-90	1987	f1
Garner and Pandy	Flexors and extensors	26 muscle groups	33	2003	f2
Koo and Mak	Flexors and extensors	BIC, BRA BRD, Triceps	100	2005	f3
Holzbaur et	Flexors and extensors	Forearm and hand	45	2005	f4
	Flexors and extensors	Elbow and showlder	140		

Table 4.5 The value of σ employed in simulation study.

Note that all the values appeared in each article are not shown in Table 4.1~4.5. Usually an average value or a typical value appeared in each article is listed. Value with parenthesis shows a range that appeared in the article. The value with square bracket is calculated one from the other quantity. Blank column is the one that is not explicitly shown in the article.

The data are also shown graphically in Fig.4.8. Index shown at right column in Table 4.1~4.5 corresponds to the data shown in Fig.4.8. Standard deviation is expressed as solid vertical line with short horizontal bar. Dashed line represents a range of observed values in each work

Chaper 4 Maximum muscle stress and specific tension σ

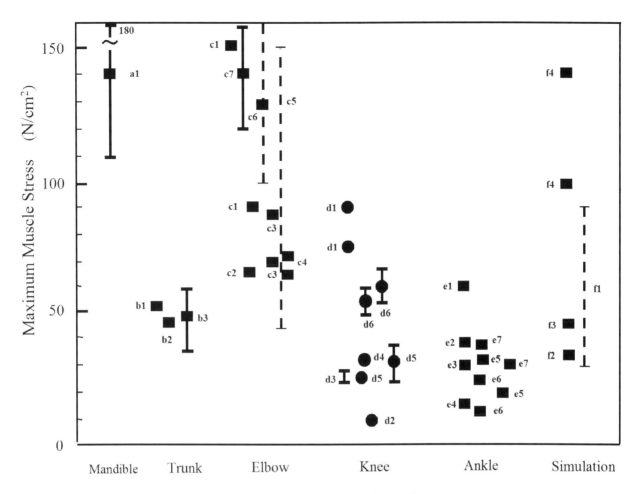

Fig.4.8 Published value of maximum muscle stress σ [N/cm²] for different muscle groups. The index a~f corresponds to the data shown in Table 4.1~4.5.

It is noticed from Fig.4.8 that value of σ is distributed in a wide range of 9~180 N/cm². The highest value was reported by Prium et al. during static bite situation for human mandible (Prium et al., 1980). They examined joint forces acting on mandible of 7 male subjects, and the highest among them was 180 N/cm². It is also noticed that reported values of each muscle group are distributed within a certain range of σ . Next highly distributed data are the ones for elbow extensors and flexors, which are in the range of 50~150 N/cm². The data of trunk and knee muscles are distributed in the range of 20~60 N/cm² except for d1 and d2. Muscles of ankle show low σ as 10~40 N/cm² except for e1. What does it mean that each muscle group shows a certain range of σ ?

It is frequently stated that the ratio of maximum muscle force per cross-sectional area converges to a single value for all of the muscle groups. It is probably based on a simple concept that basic process of interaction of Myosin and Actin filaments that exert force of muscle is the same for all animals.

Chaper 4 Maximum muscle stress and specific tension σ

The change of terminology from "Absoulute muscle force" to "Specific tension" for σ				
Author	Muscle groups	Name	Usage	Year
Hermann	Plantar flexors	Force per area		1898
Reys	Human trunk	Absolute muscle force per cross section		1915
Haxton	Plantar flexors	Absolute muscle force	Specific	1944
Morris	Elbow flexors/ Knee flexors	Strength of muscle per cross section	Common	1948
Ikai and Fukunaga	Elbow flexors	Muscle strength per cross-sectional area	Common	1968
Prium et al.	Human mandible	Maximum muscle tension	Specific	1980
Hatze	Elbow extensors	Absolute muscular forces per area		1981
Shultz et al.	Human trunk	Largest required mean muscle contraction intensity	Specific	1982
Wickiewicz	Plantar flexors	Maximum force per PCSA	Specific	1984
Vandervoort and McComas	Plantar flexors	Maximum voluntary strength	Specific	1986
Edgerton et al.	Elbow flexors	Specific tension	Specific	1986
Davies et al.	Plantar flexors	Specific tension	Specific	1986
Reid and Costigan	Human trunk	muscular strength per unit cross-sectional area	Specific	1987
Narici et al.	Knee extensors	Muscle stress	Specific	1992
Kawakami	Elbow flexors	Specific tension	Specific	1994
Akima et al.	Knee extensors	Specific tension	Specific	1994
Buchanan	Elbow flexors	Maximum muscle stress and specific tension	Q. for common	1995
Fukunaga et al.	Plantar flexors	Specific tension	Q. for common	1996
Koo et al.	Elbow flexors	Maximum muscle stress	Specific	2002
Reeves et al.	Knee extensors	Muscle specific force	Specific	2004
Langenderfer et al.	Elbow flexors	Specific tension	Specific	2005
O'Brien et al.	Knee extensors	Specific tension	Specific	2009

Table 4.6 How was the parameter σ called in the literature. The column name represents terminology used in the reference. The column denoted "usage" means how the constant σ is considered. "Specific" means σ is treated as muscle specific stress. "Common" means σ is considered as common constant for all muscles. "Q for common" expresses that the author of the paper doubted on constant value of σ for all muscles. It can not be judged for the usage of σ in the work expressed as blank column.

Chapter 4 Maximum muscle stress and specific tension σ

It is actually true, however, there are many process involved in transferring force from molecule level to individual behavior level. And, the mechanism involved in transferring force is different for muscle groups. As a result, the maximum muscle stress σ shows a different value for each muscle group. Then, a term "specific tension" became to be used for expressing muscle specific maximum stress among muscle physiologists from mid 1980's. As far as the author concern, researchers who used this term at first were Edgerton et al. (Edgerton et al., 1986) and Davies et al. (Davies et al., 1986). Table 4.6 shows a list how did it be called in the literature.

It is noticed from Table 4.6 that the constant σ has been called as strength or force of muscle per cross-sectional area, and so forth until early 1980's. Whatever it was called, the value was considered to be the same for all muscles until late 1960's. From early 1980's, it was recognized to be different according to the difference of muscle groups. Then, the constant was treated as muscle specific quantity in the works (Prium et al., 1980; Shultz et al., 1982). Thus, from around 1980's muscle physiologists became to consider that value of the ratio of force per cross-sectional area was not the same for all muscles, but muscle specific one. Then, from the latter half of 1980's, σ has been called as specific tension except for some exception.

How about maximum tetanic tension P_o ? Because the measurement of this quantity does not have long history as σ, the researchers did not seem to be thought it to be the same for all species of animals. Rather, it was noticed from early times of the research history that characteristics of muscle fibres were not the same for all muscles. Jahromi and Atwood stated as follows in 1969;

"recent work on numerous muscles has shown that muscle fibers can differ in a number of structural and biochemical features besides sarcomere length", (Jahromi and Atwood, 1969).

They noticed that the deep extensor muscles of lobster (*Homarus americanus*) contracted faster than superficial extensor muscles, and P_o of those were different values. The reported that the value of P_o is 45.15 ± 7.56 (S.E.) N/cm^2 for slow muscles and 8.38 ± 1.49 (S.E.) N/cm^2 for fast muscles. In 1985 Johnston discussed on different characteristics of contractile properties in different fibre types; pale/pink multiply innervated, red multiply innervated, pink/red focally innervated and mammalian slow twitch muscles (Johnston, 1985). Thus, from an early stage of muscle contractile study it was considered that P_o was not universal constant for all muscle types.

Chaper 4 Maximum muscle stress and specific tension σ

Common Name	Species	Muscle	P_0 [N/cm^2]	Author	Year
Crustaceans					
Crab	*Cancer branneri*	claw closer	103.2 ±3.5 (SEM) (range=71.3-153.6)	Taylor	2000
	Cancer oregonensis	claw closer	100.7 ±3.0 (SEM) (range=81.7-134.6)	Taylor	2000
	Cancer antennarius	claw closer	86.6 ±6.2 (SEM) (range=55.1-118.2)	Taylor	2000
	Menippe mercenaria	claw closer	220.0	Blundon	1988
	Callinectes sapidus	claw closer (crusher)	63.8 ±17.8 (SD)	Govind et al.	1985
Lobster		slow abdominal	45.11 ±7.56 (SEM)	Jahromi et al.	1969
		claw closer	39.0	Elner et al.	1981
Mollusks					
Squid	*Loligo pealei*	transverse (of arm)	46.83 ±9.12 (SD)	Kier et al.	2002
blue mussel	*Mytilus edulis*	anterior byssus retract	55.0 ±4.0 (SD)	Gilbert	1978
fan mussel	*Pinna fragilis*	posterior adductor	58.8	Abbott et al.	1956

Table 4.7 Experimental results which show large P_0 are excerpted from Table 3.1. Taylor's results are added from this article.

Especially, it was well known that P_o of crustaceans and molluscs was far larger than the ones of the other groups. These are listed in Table 4.7. P_o of amphibians were also comparatively larger than the other groups such as reptiles, fish, insects and mammals.

Figure 4.9 shows the data of P_o of crustaceans and mollusks together with the other animal groups, which is a graphical version of Table 3.1. It is obviously noticed that P_o of crustaceans and mollusks is significantly higher than the other groups. Among them, the maximum value of P_o was reported by Blundon (Blundon, 1988), that is P_o=220 N/cm² for claw closer muscle of crab (*Menippe mercenaria*).

Chapter 4 Maximum muscle stress and specific tension σ

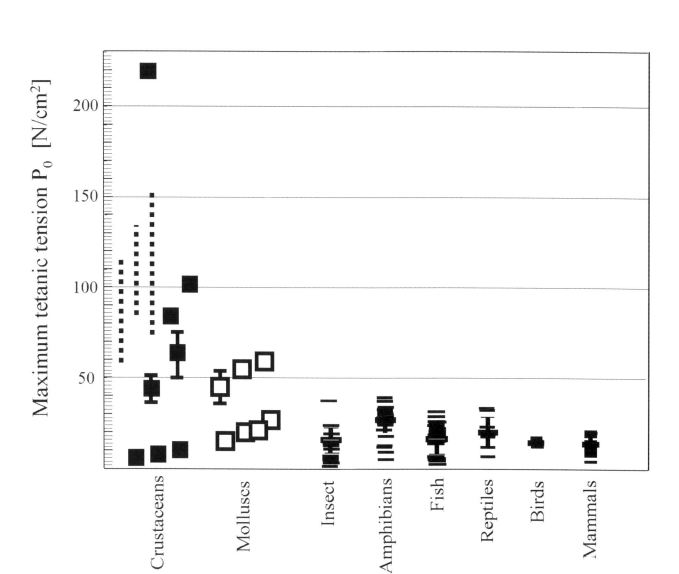

Fig.4.9 Maximum tetanic tension P_0 of crustaceans and molluscs together with the other animal groups. It is obvious that P_0's of crustaceans and molluscs show very large values compared to the ones of the other animal groups.

Interestingly, the largest value of observed specific tension σ shows similar absolute value of P_0 as σ =180 N/cm² for human mandible recorded during a bilateral static bite (Prium et al., 1980). Comparatively similar large value of σ was also reported as σ =153 N/cm² for elbow extensors

Chaper 4 Maximum muscle stress and specific tension σ

(Morris, 1948), σ =148 N/cm² for elbow flexors (Buchanan, 1995), and σ =141±20 N/cm² for elbow flexors (Langenderfer et al., 2005).

Now, there are two ways of interpretation of these experimental data for large specific tension σ.

1 Interpret these as some kind of error in the experiments.

   True value of σ is lower than these observed value (140~180 N/cm²).

2 Muscle actually exerts such high force per cross-sectional area.

It is said that taking the interpretation 2 is suitable than the interpretation 1. One reason is that there were several experimental results, not one, that reported such large value of σ. Those were obtained by different experimental methods. Next, there is no uniform tendency that value of σ converges to a small value or large value as time passes. In different time and in different experimental setups such large value of σ has been reported. Then, it would be natural to consider that muscle actually exerts such high force per cross-sectional area. If reported value reflects the true value of , some mechanism must be involved in transferring force from molecule level to muscle level of individual that suppresses final muscle force.

A possible interpretation of this phenomenon is to consider that these are the result of adaptation in evolution occurred for each species and for each muscle group. Figure 4.10 expresses schematic representation of relative strength of P_o and σ. Sliding motion of Myosin and Actin filaments leads to shortening of muscle fibre. The sum of all the shortening force minus passive resistance yields to the theoretical maximum of P_o. However, the process of transmission of the force between them has not been resolved. Then, theoretical maximum value of P_o is not known yet. Observed maximum value of P_o is 220 N/cm², which was found for crustacean (crab, *Menippe mercenaria*).

Chapter 4 Maximum muscle stress and specific tension σ

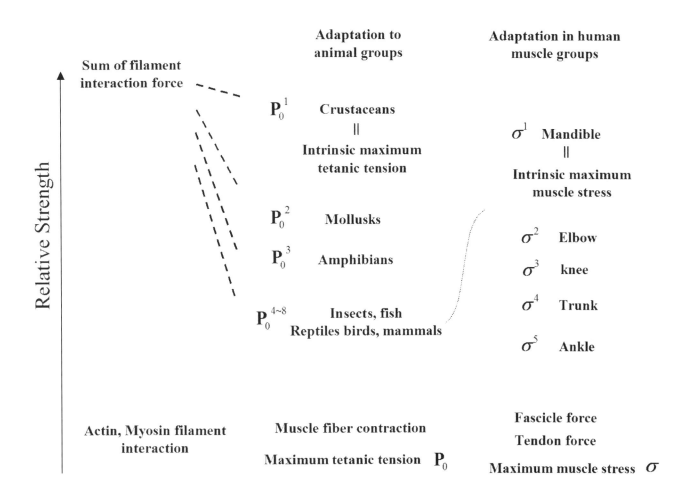

Fig.4.10 Schematic representation of relative strength of P_0 for different animal groups and σ of different human muscle groups. Value of maximum tetanic tension P_0 is not the same for different animal groups. Similarly, maximum muscle stress σ is also not the same for different human muscle groups.

Let denote it as P_0^1. Maximum tetanic tension for the other animal groups is smaller than this. Then, observed value of P_0 for those animal groups could not be said as intrinsic maximum tetanic tension. It is said to be adapted muscle strength for each animal group, that is fitted for those living habits. There is no common value of P_0 applied for all animals, but should be labeled as $P_0^2, P_0^3, P_0^4, \cdots$.

60

Chapter 4 Maximum muscle stress and specific tension σ

The situation is similar for the maximum muscle stress σ. Prium et al reported that a subject marked σ =180 N/cm² for human mandible. This should be assigned as intrinsic maximum muscle stress. Muscle can exert force up to this strength at best. Let denote this as σ^1. It is considered that adaptation of the system within and without muscle has been occurred through evolution, Then, each muscle system has been fitted to have lower strength of muscle force than σ^1. Fukunaga et al. reported a low value of σ for plantar flexors as 11 N/cm² (Fukunaga et al., 1996). They stated about factors reducing force as follows;

"a considerable amount of force generated in a muscle fibre can be transmitted via the inter fibre matrix, for ankle joint plantar flexors and dorsiflexors extend only a portion of the muscle length".

Factors outside muscle are morphological and functional differences of attachment from muscle to bone through tendon. Then, when we observe maximum muscle stress, different values of σ are obtained, those may be denoted as $\sigma^2, \sigma^3, \sigma^4, \cdots$. Thus, different values of σ have been reported for different muscle groups, and most of those correctly describe strength of muscle for each muscle groups.

We have to pay attention to the followings. Index of muscle strength has been frequently reported as lower value than adequately measured. O'brien et al. stated as follows,

"specific tension in that study was underestimated because muscle forces were not maximal, since the muscles studied were activated by percutaneous electrical stimulation, which is unlikely to recruit the whole muscle", (O'brien et al. 2009).

We have to consider that those do not describe correctly intrinsic muscle strength of each muscle group.

The following should also be remarked that a wide range of σ is observed in each experimental report depending on subjects. For example, scattered value of force and cross-sectional area are observed in Fig.4.3, the value of σ is 66±11 N/cm². It is difficult to consider that all data express maximum muscle stress or specific tension of muscle. But, it is better to consider as follows. The value of intrinsic maximum muscle stress is 180 N/cm² which was observed for human mandible. The others are not the intrinsic maximum muscle stress, but muscle stress adapted for each muscle group.

Finally, it is worthwhile to note the following. In these years, we could use MRI scanning device. Then, the quantity that the force per cross-sectional area may not be a good measure to express muscle

Chaper 4 Maximum muscle stress and specific tension σ

stress. It is an idea that proposed a century and one half ago. Now, we are at an age that muscle is measured as free-form object, not a cylinder. Then each quantity such as fibre length, cross-sectional area are re-calculated from MRI images. A better index which expresses muscle strength may be exist in modern muscle physiology. One alternative is to consider the relation that muscle volume v.s. joint torque. This idea was proposed by Fukunaga et al. By using this relation, there is no need to re-calculate average value of the quantities such as fibre length and cross-sectional area. This relation is discussed in detail in the next chapter.

For closing this section, the followings are summarized.

1. Until 1960's, it was considered that there was a common maximum muscle stress which was applied for all muscles.

2. From 1980's, it has been considered that a force per cross-sectional area depends on muscle groups. At present, the constant σ is called specific tension.

3. A variance of observed value of specific tension is large, which span in a range of 9~180 N/cm². One possible interpretation of it is to consider that those are results of adaptation for species, muscle groups and subjects.

4. One experiment showed that the maximum of σ was 180 N/cm², and several experiment showed that the maximum of σ was 140 N/cm². These values are considered as the maximum force that muscle exerts at best per its cross-sectional area.

5. A situation is similar to the maximum tetanic tension P_0. The maximum value of P_0 was 220 N/cm². Distributed values within the range of 11~220 N/cm² are considered to be a result of adaptation that fits for the species, muscle groups and subjects.

Concerning to muscle of dinosaurs, the following is said. Apparently, dinosaurs are not within known extant animal groups. Then, there should be no logical criteria that correctly estimate muscle stress of dinosaurs. For an estimation, if we employ σ of human specific tension, $\sigma = 60$ for knee or $\sigma = 40$ for ankle may be a better choice. However, we can not exclude the possibility that σ is larger than these values. The observed maximum of σ was 140 N/cm², which was found for human elbow by several research groups, not one. Then, possibility that σ of dinosaur leg is larger than $\sigma = 60$ N/cm² can not be excluded.

Chaper 4 Maximum muscle stress and specific tension σ

References

Amis, A., A., Dowson, D., and Wright, V. Muscle strengths and musculo-skeletal geometry of the upper limb. Eng. Med. (1979) 8: 41-48.

Arkin, A. M. Absolute muscle power. The internal kinesiology of muscle. Arch. Surg. (1941) 42(2): 395-410(1941).

Buchannan, T. S., Evidence that maximum muscle stress is not a constant: differences in specific tension in elbow flexors and extensors, Med. Eng. & Phys. (1995) 17: 529-536.

Davies, C. T. M., Thomas, D. O. & White, M. J. Mechanical properties of young and elderly human muscle. Acta Med. Scand. -Suppl. (1986) 711: 219-226.

Edgerton, V., R., Roy, R., R., and Apor, P. Specific tension of human elbow flexor muscles. In: Saltin B (ed) Biochemistry of exercise VI. Human Kinetics, Champaign, I11. (1986): 487-500.

Edgerton, V. R., Apor, P., and Roy, R. R., Specific tension of human elbow flexor muscles, Acta Phyiol. Hung. (1990)85: 205-216.

Fick, R., Handbuch der Anatomie und Mechanik der Gelenke, Fischer Verlag, Jena (1904) : 530.

Fukunaga, T., Roy, R. R., Shellock, F.G., Hodgson, J. A., and Edgerton, V.R., Specific tension of human plantar flexors and dorsiflexors. Amer. Physiol. Soc. (1986) 80: 158-165.

Garner, B. A. and Pandy, M.G., Estimation of musculotendon properties in the human upper limb. Ann. Biomed. Eng. (2003) 31: 207-220.

Hatze, H., Estimation of myodynamics parameter values from observations on isometrically contracting muscle groups, Eur. J. Appl. Physiol. O. (1981) 46: 325-338.

Haxton, H. A., Absolute muscle force in the ankle flexors of man, J. Appl. Physiol. (1994) 103: 267-273.

Hermann, L., Zur messung der Muskelcraft am Menschen, Pflugers Archiv (1989)73: 429-437.

Hertinger, T. in Physiology of Strength (Thurlwell, M.H. ed.) (Springfield, Thomas, IL., 1964).

Holzbaur, K.R.S., Murray, W.M. & Delp, S.L. A model of the upper extremity for simulating musculoskeletal surgery and analyzing neuromuscular control. Ann.Biomed. Eng. (2005) 33: 829-840.

Ikai, M. and Fukunaga, T., Calculation of muscle strength per unit cross-sectional area of human muscle by means of ultrasonic measurement. Int Z Angew Physiol. (1968) 26: 26-32.

Jahromi, S. S. and Atood, H. L., Correlation of Structure, Speed of Contraction and Total Tension in Fast and Slow Abdominal muscle fibers of the lobster, J. Exp. Zool. (1969) 171 : 25-38.

Johnson, J. V. Ergebnisse der Physiologie (1903) 2(2): 623.

Johnston, I. A., Austained force development: specializations and variation among the vertebrates, J. exp. biol. (1985) 115 : 239-251.

Kawakami, Y., Nakazawa, K., Fujimoto, T., Nozaki, D., Miyashita, M., and Fukunaga, T., Specific tension of elbow flexor and extensor muscles based on magnetic resonance imaging. Eur. J. Appl. Physiol. (1994) 68: 139-147.

Koo, T. K. K. and Mak, A. F. T. Feasibility of using emg driven neuromuscu-loskeletal model for prediction of dynamic movement of the elbow. J. Electromyogr. Kinesiol. (2005) 15: 12-26.

Koo, T. K. K. Mak, A. F. T., and Hung, L. K. In vivo determination of subject-specific musculotendon parameters: applications to the prime elbow flexors in normal and hemiparetic subjects. Clin. Biomech. (2002) 17: 390-399.

Langenderfer, J. Jerabek, S.A., Thangamani, V.B., Kuhn, J.E., and Hughes, R.E. Musculoskeletal parameters of muscles crossing the shoulder and elbow and the effect of sarcomere length sample size on estimation of optimal muscle length. Comp. Biol. Med. (2005)35: 25-39.

Maganaris, C. N., Baltzopoulos, V., Ball, D, and Sargeant, A. J. In vivo specific tension of human skeletal muscle. J. Appl. Physiol. (2001) 90: 865-872.

Maughan, R., J., Watson, J., S., and Weir, J. Strength and cross-sectional area of human skeletal muscle, J. Physiol. (1938) 338 : 37-49.

McGill, S. M. and Norman, R. W. Effects of an anatomically detailed erector spinae model on L4/L5 disc compression and shear. J. Biomech. (1987) 20(6) : 591-600.

Morris, G. B. The measurement of the strength of muscle relative to the cross-section. Res. Quart. Amer. Assoc. Health, Phys. Edu. (1948) 19: 295-303.

O'Brien, T. D., Reeves, N. D., Baltzopoulos, V., Jones, D. A., and Maganaris, C.N., *In vivo* measurements of muscle specific tension in adults and children, Exp. Physiol. (2009) 95:202-210.

Prium, G. T., de Jongh, H. J., and ten Bosch, J. J. Forces acting on the mandible during bilateral static bite at different bite force levels, J. Biomech. (1980) 13: 55-63.

Rechlinghausen, N. Gliedermechanik und Lahmungsprothesen (Springer, Berlin, 1920).

Reeves, N. D., Narici, M. V., and Magnaris, C. N., Effect of resistance training on skeletal muscle-specific force in elderly humans, J. Appl. Physiol. (1985) 96: 885-892.

Reid, J. J. G. and Costigan, P. P. A., Trunk muscle balance and muscular force. Spine (1987) 12: 783-786.

Reys, J. H. O., Uber die absolute muskelkraft in menschlichen korper. Pflug. Arch. Eur. J. Phy. (1915) 160: 133-204.

Weber, E. 1846. Wagner's Handworterbuch der physiologie. Braunschweig, Vieweg.

Chaper 4 Maximum muscle stress and specific tension σ

Appendix

This subsection describes consistency of notation presented so far and the reference (Hutchinson and Garcia, 2002). Eq.(4.2) is written as,

$$\text{PCSA} \cdot l_f = V \cos\theta \quad . \tag{A4.1}$$

The definition of maximum muscle stress σ is,

$$\sigma = \frac{F_t}{\text{PCSA}} \quad . \tag{A4.2}$$

These lead to the following,

$$V = \frac{F_t \cdot l_f}{\sigma \cdot \cos\theta} \quad . \tag{A4.3}$$

Let us write i-th joint mass as m_i. Then, the volume V is interpreted to corresponding volume for m_i. We have a relation of $m_i = V \cdot d$ with density d.

Then, Eq.(A4.3) becomes,

$$\frac{m_i}{d} = \frac{F_t \cdot l_f}{\sigma \cdot \cos\theta} \quad . \tag{A4.4}$$

The moment of force M_i is defined by a product of force F_i and moment arm r,

$$M_i = F_t \cdot r \quad . \tag{A4.5}$$

This leads to,

Chaper 4 Maximum muscle stress and specific tension σ

$$m_i = \frac{M_i \cdot l_f \cdot d}{r \cdot \sigma \cdot \cos\theta} \quad . \tag{A4.6}$$

Then, we have a percentage of body mass $m_i(\%)$,

$$m_i(\%) = 100 \cdot \frac{1}{m_{body}} \frac{M_i \cdot l_f \cdot d}{r \cdot \sigma \cdot \cos\theta} \quad . \tag{A4.7}$$

In the notation appeared in the article (Hutchinson and Garcia, 2002), $m_i(\%)$ is expressed as,

$$m_i(\%) = \frac{100 M_i L d}{\sigma c r m_{body} \cos\theta} \quad . \tag{A4.8}$$

In the expression each parameters and the values are the following,

c	the fraction of active muscle volume	c=1
d	the muscle density	$d = 1.06 \times 10^3$ kg/m³
m_{body}	the whole *T.rex* body mass	$m_{body} \fallingdotseq 6000$ kg

Eq.(3.14) becomes more simple form as,

$$m_i(\%) = \alpha \cdot \frac{M_i L}{r \cos\theta} \quad . \tag{A4.9}$$

with a set of known parameters $\quad \alpha = \dfrac{100 \cdot d}{\sigma \cdot c \cdot m_{body}} \quad . \tag{A4.10}$

Chapter 5 Torque v.s. Muscle volume

In the previous chapter, reported values of the maximum muscle stress σ (the specific tension) show a large variance from 9 N/cm²~180 N/cm². Many researchers have warned to use it for the quantitative estimation of muscle force from the physiological data.

Instead of using the specific tension σ, Fukunaga et al. suggested the use of muscle volume for the index of muscle strength (Fukunaga et al., 2001). This idea does not need such quantity as physiological cross section (PCSA) and moment arm. At present, we can measure the shape of muscle directly *in situ* using MRI, etc. Then, the quantities PCSA and moment arm may become old fashioned for describing muscle properties, because the each one is the quantity of averaged value over the muscle. Fukunaga et al. showed that the joint torque has a linear relation to the muscle volume as well as the relation of force and PCSA. As a bonus, there is no need to introduce PCSA and moment arm in the relation of the joint torque and the muscle volume. Especially, it should be noted that PCSA is not the quantity that measured directly, but calculated as an average value from muscle volume and fibre lengths. Then, using the ratio of the torque per the muscle volume may be appropriate for expressing the muscle strength. Akagi et al.'s following work in 2009 is introduced in this chapter (Akagi et al, 2009).

In 2009 Akagi et al. published results of more detailed study concerning to the relation of torque v.s. muscle volume (Akagi et al., 2009). They examined which between the muscle volume or cross-sectional area has an appropriate relation with the elbow flexor muscle strength. The 103 subjects of young and old, man and woman were examined. The muscle volume and the anatomical cross-sectional area of elbow flexors were determined by the measurements of magnetic resonance image (MRI). The torque and the force were determined by the maximal voluntary contraction of isometric elbow joint flexion. A detailed description of the method is given in the reference (Akagi et al. 2009), and references therein. Figure 5.1 contains data of the followings;

(a) Torque v.s. muscle volume for young and old men
(b) Torque v.s. muscle volume for young and old women
(c) Force v.s. ACSA for young and old men
(d) Force v.s. ACSA for young and old women

Chapter 5 Torque v.s. Muscle volume

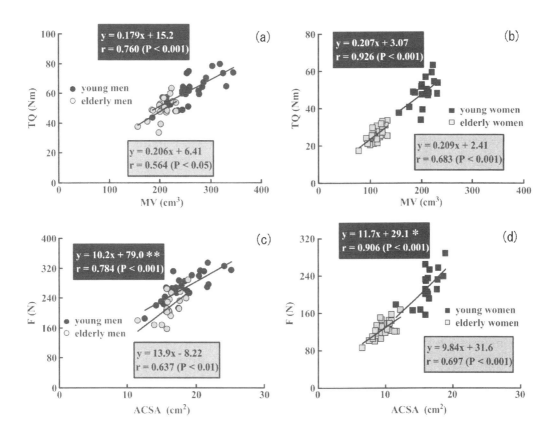

Fig.5.1 Experimental results for elbow flexor muscles of 103 subjects, (a) torque v.s. muscle volume for men (young and old), (b) torque v.s. muscle volume for women (young and old), (c) force v.s. ACSA (anatomical cross-sectional area) for men, (d) force v.s. ACSA for women, (Akagi et al. 2009).

For comparing men's data i.e., (a) and (c), toruque (TQ) v.s. muscle volume (MV) and force (F) v.s. anatomical cross sectional-area (ACSA) are similarly correlated in r and P. The r expresses correlation coefficient of two variables, which has a value in the range of -1 to 1. If r is 1, two variables have the maximum positive correlation. The P expresses significant level. If P<0.001, there is a 0.1 percent chance that the result was accidental.

Young men
Torque (y) v.s. Muscle volume (x) $y = 0.179\,x + 15.2$ (r=0.760   P<0.001)
Force (y) v.s. Anatomical cross-sectional area (x) $y = 10.2\,x + 79.0$ (r=0.784   P<0.001)

The r values are similar in the both correlations. However, Akagi et al. pointed out that y−intercepts of the regression lines are different as 15.2 for TQ v.s. MV and 79.0 for F v.s. ACSA. Then, they argued that the

Chapter 5 Torque v.s. Muscle volume

relation of torque v.s. muscle volume is a better indicator to express muscle strength than the relation of force v.s. anatomical cross-sectional area.

Let us check on this point in each correlation. The regression relations are,

Torque (y) v.s. Muscle volume (x)

$y = 0.179\,x + \underline{15.2}$ (young men)
$y = 0.206\,x + \underline{6.41}$ (elderly men)
$y = 0.207\,x + \underline{3.07}$ (young women)
$y = 0.209\,x + \underline{2.41}$ (elderly women)

Force (y) v.s. Anatomical cross-sectional area (x)

$y = 10.2\,x + \underline{79.0}$ (young men)
$y = 13.9\,x - \underline{8.22}$ (elderly men)
$y = 11.7\,x + \underline{29.1}$ (young women)
$y = 9.84\,x + \underline{31.6}$ (elderly women)

In principle, if there is no muscle, there must be no force. Bruce et al. also reported that F v.s. ACSA cannot have a true intercept, namely, y (force) $\neq 0$ when x (ACSA)=0 (Bruce et al., 1989). In the above statistics, the relation of force v.s. anatomical cross-sectional area yields large y-intercept. The exception is the case for elderly men in the relation of force v.s. ACSA.

Aging effect

Akagi et al. reported that aging effect is observed in F/ACSA, however, it is not significant in TQ/MV. Table 5.1 shows the data concerning to aging effect.

	Men			Women		
	Young	Elderly	Young/Elderly	Young	Elderly	Young/Elderly
Torque/Muscle volume	23.9±2.4	23.7±0.7	1.01	23.0±1.5	23.0±2.5	0.00
Force/ACSA	14.7±1.4	13.4±1.7	1.10	14.7±1.3	13.1±1.3	1.12

Table 5.1   Aging effect in the relation of Torque/Muscle volume and Force/ACSA.
           The author calculates the ratio of young per elderly for each mean value.

Chapter 5 Torque v.s. Muscle volume

The authors calculated the ratio of young per elderly for the mean values of men's and women's data. From the data, it is suggested that the ratio of force per ACSA shows relatively large values for both of men and women. On the contrary, it is not apparent to observe aging effect for the ratio of torque per muscle volume, for both of men and women. Then, from this viewpoint it is stated that the ratio of torque per muscle volume is appropriate for expressing muscle strength rather than the ratio of force per ACSA.

In concluding this chapter, it should be noted that this work was accomplished for elbow flexor muscle. Further studies of the other muscle groups are required to make the statement for concrete. Based on the works (Akagi et al., 2009; Fukunaga et al., 2001), the followings are given as the summary of this chapter;

- Large y-intercept is observed in the regression line of F (Force) v.s. ACSA (anatomical cross sectional-area). In principle, F should be zero when ACSA is zero. On the contrary, torque is nearly zero when the muscle volume is zero in the statistics.

- Aging effect is observed in F/ACSA.

- A significant difference for the different age group is not observed in TQ/VM.

- Then, $\dfrac{\text{Torque}}{\text{Muscle volume}}$ is a better indicator to express muscle strength than $\dfrac{\text{Force}}{\text{ACSA}}$.

References

Akagi, R., Takai, Y., Ohta, M., Kanehisa, H., Kawakami, Y., and Fukunaga T., Muscle volume compared to cross-sectional area is more appropriate for evaluating muscle strength in young and elderly individuals. Age Ageing. (2009) Sep;38(5):564-9. doi: 10.1093/ageing/afp122. Epub 2009 Jul 13.

Bruce, S., A., Newton, D., and Woledge, R., C., Effect of age on voluntary force and cross-sectional area of human adductor pollicis muscle. Q J. Exp. Physiol. (1989) 74: 359–62.

Fukunaga, T., Kubo, K., Kawakami, Y., Fkashiro, S., Kanehisa, H., and Maganaris, C., N., In vivo behaviour of human muscle tendon during walking Proc. R. Soc. Lond. B. (2001) 268: 229-233.

Chapter 6 Mechanical Power Output of Muscle

6.1 The maximum isotonic power

The maximum isotonic power is a quantity to measure mechanical power of muscle with the condition of constant load. In physics the work W is defined as contour integral of the product of the force $\vec{F}$ and the path $\vec{x}$,

$$W = \oint \vec{F} \cdot d\vec{x} \qquad . \tag{6.1}$$

This is equal to the area surrounded by a closed loop in the diagram of the force and the path for one dimension system. Consider a case for the space dimension as being one. And if F and x are constant, cyclic motion becomes rectangular as shown in Fig.6.1. Then, Eq.6.1 becomes,

$$W = F_1 \vec{x}_1 + F_2 \vec{x}_2 \tag{6.2}$$

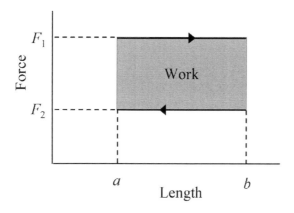

Fig.6.1 The area $(F_1 - F_2)(b - a)$ corresponds to the mechanical work in the diagram of length v.s. force

Note that $\vec{x}_1 = b - a$, $\vec{x}_2 = a - b = -(b - a)$. Then, Eq.6.1 becomes,

$$W = (F_1 - F_2)(b - a) \qquad . \tag{6.3}$$

This corresponds to the area surrounded by the closed loop in Fig.6.1.

Chapter 6   Mechanical Power Output of Muscle

6.2 Work Loop Technique

Muscle contraction in nature does not always occur in a condition of constant load as introduced in Chap.3. Since the publication of Hill's work[15], numerous experimental results had been published concerning to isometric muscle contraction. (The reference number is listed in Table 6.1 and Fig.6.6, 6.7.) On the contrary, work production by muscle had been infrequently studied and relatively poorly characterized, which was Josephson's indication. In 1985 Josephson proposed work loop technique to study power output from striated muscle during cyclic contraction. Instead of giving constant load to the muscle, series of stimulus simulating such as wing beat of birds are given to the muscles. Josephson stated on the situation at that time as follows;

"But muscles in vivo rarely if ever operate at constant load, so isotonic measurements are not those of normal operating conditions. Further, most isotonic work studies have not considered the work which must be done on muscle after it has shortened to restretch it to its original length, a factor whose magnitude may be appreciable at higher operating frequencies. (Josephson, 1985) [16]"

Note that work loop technique was described by Josephson in 1985, however, he stated that prior works concerning this were published by Machin & Pringle in 1960[17], Jewell & Rüegg in 1966[18], Steiger & Rüegg in 1969[19], and Kawai & Brandt in 1980[20]. After Josephson's publication, studies on work output of muscle have been increased, and understanding on muscle movement properties has been progressed.   Especially, comparison of mechanical work output produced at the level of bundles of fibres, and the level of individual movement has been emerged. This comparison brought much knowledge to muscle physiology, and at the same time new unsolved questions have been raised.

When a muscle moves cyclically, periodical movement of muscle length and force in time will be obtained. An example of those is shown in Fig.6.2(A). If we plot the data as length v.s. force, closed loop would be obtained, which is shown in Fgi.6.2(B).

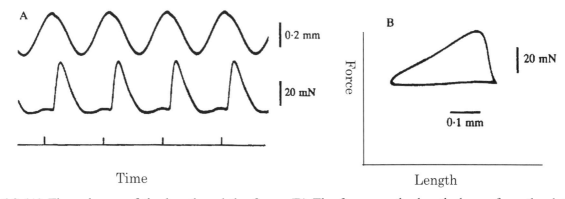

Fig.6.2 (A) Time change of the length and the force. (B) The force v.s. the length drawn from the data (A),

Chapter 6 Mechanical Power Output of Muscle

(adapted with permission from Fig.1 of The Journal of Experimental Biology, Josephson, 1985, vol.114, pp493-512)

Muscle produces shortening force only. No lengthening force of muscle has yet been observed. Then, positive work is defined by the mode of motion as follows;

- **Positive work**   If a muscle shortening occurs, the muscle would do the work on the limb. This movement shows counterclockwise loop in force-length diagram. This movement is called positive work (Fig.6.3).

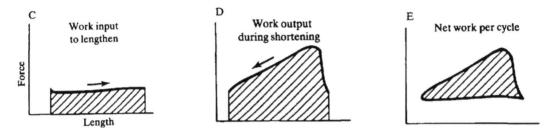

Fig.6.3   Positive work is defined as counterclockwise loop in force-length diagram. (Adapted with permission from Fig.1 of The Journal of Experimental Biology, Josephson, 1985, vol.114, pp493-512)

On the contrary, negative work is defined as follows;

- **Negative work**   If a muscle lengthens at higher force, and shortens at lower force, The movement is called negative work. This situation occurs in such a case of constant- speed downhill walking.

6.3 Experiment using work loop technique

Askew and Marsh's study for the pectoralis muscle of blue-breasted quail is introduced as an example of an experiment using work loop technique (Askew and Marsh, 2001)[12]. They achieved analysis for the muscle *in vivo* and *in vitro*. Sonomicrometry and electromyographic recordings were made for the pectoralis muscle of blue-breasted quail (Coturnix chinensis) during take-off. Then, strain activity *in vivo* were measured. Next, bundles of fibres were dissected from the pectoralis and subjected in vitro to the in vivo length and activity patterns, whilst measuring force. Consequently, they reported that the net power output during shortening averaged over the entire cycle was 349 W/kg as muscle-mass specified work. The peak instantaneous power output was 1121 W/kg.

73

Chapter 6　Mechanical Power Output of Muscle

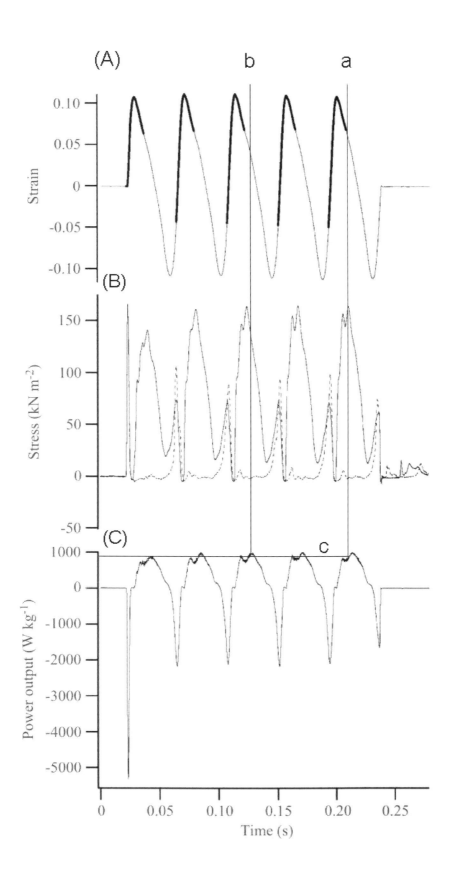

Chapter 6 Mechanical Power Output of Muscle

Fig.6.4 An example of data obtained from the *in vitro* work loop experiments for the bundles of fibres isolated from the quail pectoralis, (Askew and Marsh, 2001). The strain trajectory was simulated to express the strain measured *in vivo* using sonomicrometry (A). The muscle was activated 7 ms before peak length for 14 ms as indicated by the bold lines in (A). The (B) and (C) show stress [kN/m^2] and power output [W/kg], respectively. The instantaneous power output was calculated by the product of force times the velocity.   (Adapted with permission from Fig.5 of The Journal of Experimental Biology, Askew and Marsh., 2001, vol.204, pp3587-3600).

Fig 6.4 shows examples of data obtained from the *in vitro* work loop experiments[12]. A bundle of muscle fibres was isolated and subjected to the cyclical length changes measured in vivo using sonomicrometry. The muscle was activated 7 ms before peak length for 14 ms as indicated by the bold lines in A. Force was measured, and this has been used to calculate muscle stress as in B. The instantaneous power output was calculated by multiplying force by the velocity of shortening as C. Finally, work loops were generated by plotting the stress against strain as Fig.6.5. The arrows indicate the direction of the loop.

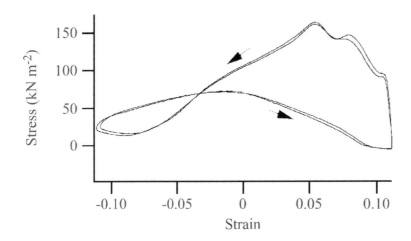

Fig.6.5 Work loops were generated by plotting the stress against strain[12]. The arrows indicate the direction of the loop. (Reproduced with permission from Fig.5 of The Journal of Experimental Biology, Askew and Marsh., 2001, vol.204, pp3587-3600)[12].

Note that isometric stress P_o was 13.09±0.54 N/cm² for 8 samples (n=8), where values are means ±S.E.M.

Chapter 6   Mechanical Power Output of Muscle

Istometric stress P_0 13.09 ± 0.54 (S.E.M.)   N/cm²   (130.9 ± 5.4 kN/m²)   (n=8)

Askew and Marsh stated that the value $P_0=13.0$ N/cm² was low in comparison with other measurements, for example, 26.9 N/cm² obtained for mouse limb muscles by Askew and Marsh, (Askew and Marsh, 1997)[21].

The vertical line (a) in Fig.6.4 drawn by the author of this book indicates the timing when the peak stress is observed in (B). The value of the peak stress exceeds 150 kN/m², which is larger than the isometric stress P_0 13.9 ± 5.4 N/cm² (130.9 ± 5.4 kN/m²). And, the timing almost corresponds to a time region when the peak instantaneous power is observed. The vertical line shown as (b) indicates the timing when the peak instantaneous power is observed in (C). The timing of vertical line (a) (the peak stress) and the line (b) (the peak instantaneous power) in the trajectory are almost similar, and those show 900 W/kg as indicated by horizontal line (c). Then, from this experimental result the followings are said.

1. The peak instantaneous muscle-specified power output is extremely high. Askew and Marsh reported the value as 1120.6±122.1 W/kg for seven samples of dissected bundles of muscle fibres.

2. The mean power output of the quail pectoralis muscle *in vitro* during the shortening phase of simulated flight strain trajectories was 349 W/kg.

3. The timing of the peak instantaneous power and the one of peak stress are almost the same.

4. The peak stress in shortening is almost similar to the isometric stress P_0, or slightly larger.

6.4   Can power obtained by muscle contraction studies explain animal movement ?

After piling up of muscle contraction studies from 1970's, measurements of high power output of animal movement such as flying and jumping have been achieved from 1990's. Then, the problem is whether the power obtained by muscle contraction studies explain power output obtained by animal movement. These data are summarized in Table 6.1, Fig.6.6 and Fig.6.7. For example frog jumping showed 822-1644 W/kg muscle-mass-specific power output (Peplowski and Marsh, 1997)[2]. Lacertid

Chapter 6   Mechanical Power Output of Muscle

lizard *Acanthodactylus boskianus* achieved 952 W/kg power output in the running motion (Curtin, et al., 2005)[7]. The data are summarized graphically in Fig.6.6. Measured data by the experiment of contraction studies of bundles of muscle fibres are summarized in Fig.4.1. It is remarkable that maximum isotonic contraction showed high power output as 730 W/kg for Zebra finch (Ellerby and Askew, 2007)[14]. Average values over the cycle in the work loop of muscle contraction *in vivo* were not relatively high as 470.8 W/kg for Gray jay (Jackson and Dial, 2011)[13]. However, peak instantaneous power outputs during the work loop show high values as 952 W/kg for Lacertid lizard (Curtin et al., 2005)[7] and 1120.6 W/kg for quail (Askew and Marsh, 2001)[12]. Then, two different opinions exist as interpretation of these experiments.

1 High power output of animal movement is in agreement with high peak instantaneous power output observed in muscle contraction study.

2 Power observed in muscle contraction study does not explain high power output of animal movement. Then, a mechanism of elastic storage together with muscle voluntary contraction should be involved.

Chapter 6 Mechanical Power Output of Muscle

Muscle-mass-specific power [W/Kg]	No. Sub.	Animal	Method 1	Method 2	Authors	Ref. No.
225–550	5**	Hylid frog	Jumping Performance		Marsh and John-Alder (1994)	1
822–1644	4	Frog (Isteopilus septentrionalis)	Jumping Performance		Peplowski and Marsh (1997)	2
318–747	8	Frog (Litoria nasuta)	Jumping Performance		James and Wilson (2008)	3
389.5 (mean), 531.2(Max)	6	Blue-breasted quail (Coturnix chinensis)	Take-off		Askew et al. (2001)	4
400	49*	Wild turkey	Running		Roberts and Scales (2002)	5
777.64±32.97 (mean.±SEM.)	18*	Guinea fowl (Numida meleagris)	Running		Henry et al. (2005)	6
940±104 (mean.±SE.)	5	Lacertid lizard (Acanthodactylus boskianus)	Running		Curtin et al. (2005)	7
495.0±15.0 (Ave.±SD.), 639 (Max)	15–20*	Yellow-footed rock wallabies	Jumping Performance		McGowan et al. (2005)	8
9600	4, (22*)	Toad	Tongue Projection		Lappin et al. (2006)	9
18129 (Max)	57*	Salamander	Tongue Projection		Deban et al. (2007)	10
81.1–102.3 (n=6)	6	Frog (Litoria nasuta)	Work loop	averaged over the cycle	James and Wilson (2008)	3
40.3–119.6 (n=3–7)	3–7	Cockatiel (Nymphicus hollandicus)	Work loop	averaged over the cycle	Morris and Askew (2010)	11
349.1±26.8 (mean.±SEM.,n=7), 433 (Max)	7	Quail (Coturnix chinensis)	Work loop	averaged over the cycle	Askew and Marsh (2001)	12
350.1±10.8 (mean.±SEM.), 470.8 (Max) in vivo 99*		Gray jay	Work loop (take-off)	averaged over the cycle	Jackson and Dial (2011)	13
214±20 (mean±SE.)	5	Lacertid lizard (Acanthodactylus boskianus)	Work loop		Curtin et al. (2005)	7
224 (Max.)	6	Frog (Litoria nasuta)	Work loop	Shortening phase	James and Wilson (2008)	3
952±89 (mean.±SE.)	5	Lacertid lizard (Acanthodactylus boskianus)	Work loop	Peak Instantaneous	Curtin et al. (2005)	7
1120.6±122.1 (mean±SEM.)	7	Quail (Coturnix chinensis)	Work loop	Peak Instantaneous	Askew and Marsh (2001)	12
265.2±17.9 (mean.±SEM.)	4	Frog (Osteopilus septentrionalis)	Maximum Isotonic	Hyperbolic-linear eq.	Peplowski and Marsh (1997)	2
533±118 (mean±SEM.)	12	Cockatiel (Nymphicus hollandicus)	Maximum Isotonic	Exponential linear eq.	Morris and Askew (2010)	11
730±58 (mean±SEM.)	7	Zebra finch	Maximum Isotonic	Exponential linear eq.	Ellerby and Askew (2007)	14

Table 6.1

Chapter 6 Mechanical Power Output of Muscle

Table 6.1 Reported high muscle-mass-specific powers are summarized. The data are grouped by the method.

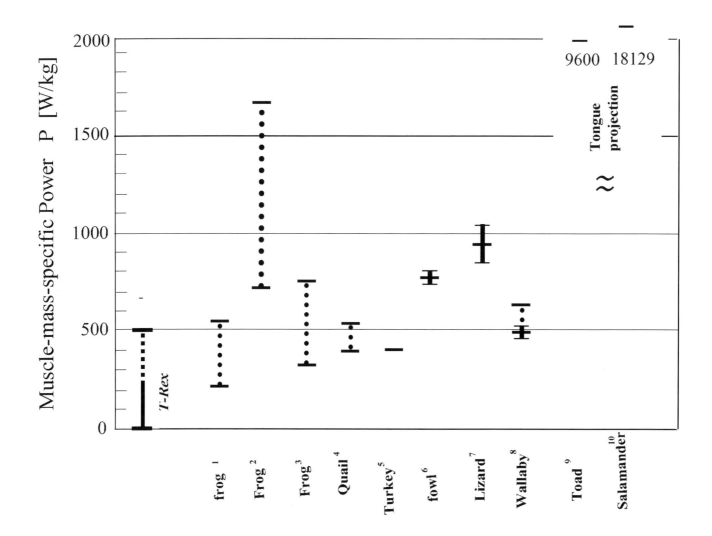

Fig.6.6 Published values of muscle-mass-specific mechanical power measured using work loop technique and isotonic contraction. The value denoted as *T-Rex* is obtained in Chap. 13.

Chapter 6 Mechanical Power Output of Muscle

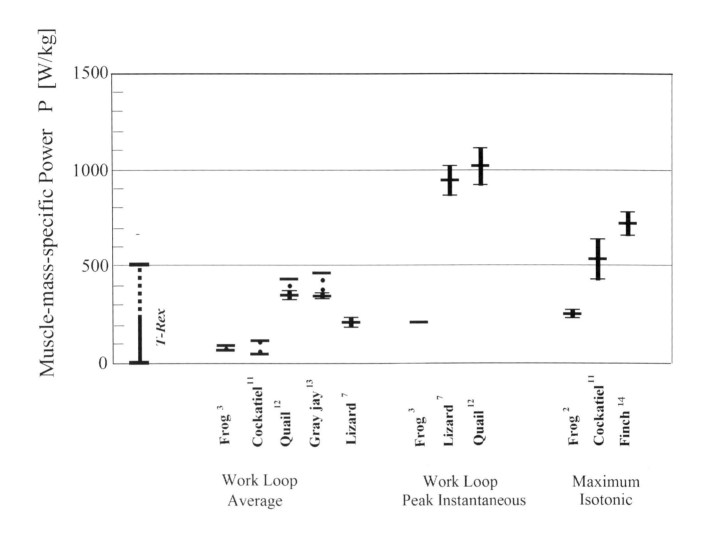

Fig.6.7. Published values of muscle-mass-specific mechanical power measured from animal movement of jumping (frog, wallaby), running (turkey, lizard), taking-off (quail, fowl) and tongue projection (toad, salamander). The value denoted as *T-Rex* is of 14.1 m/s running which will be described in Chap.8

Chapter 6 Mechanical Power Output of Muscle

Many researchers stated the possibility of No.2 interpretation, however, few stated No.1 interpretation. It is noted that extremely high power outputs are observed in the experiment of tracing movement of tongue projection of reptile. Lappin et al. reported 9,600 W/kg (Lappin et al., 2006) and Deban et al. reported 18,129 W/kg (Deban et al., 2007) for tongue projection of toad and salamander, respectively. There is no way to explain such extremely high power output from the data of muscle contraction study. Then, some kind of elastic storage should be certainly involved in the mechanism of tongue projection.

The most possible candidate is that surrounding tissue of muscle plays a role of elastic storage. Consider a spring which is connected to muscle in parallel arrangement. The spring shortens by the muscle contraction. The shortening process of the system is done slowly and by large force. As the spring start to lengthen stored kinetic energy is released. There are many works on this issue, however, those are not discussed in detail in this book. Instead, in the next chapter a role of passive element tendon is introduced in the movement of walking and jumping of human movement. In the experiment of human walking and jumping, it has been revealed that passive element tendon plays essentially important role in the movement.   As a summary of this chapter, the followings are said;

1. High power outputs are observed in the movements of animal such as jumping and taking-off. All of them could not always be explained from the data of muscle contraction study. Then, many researchers consider that the mechanism of elastic storage should be involved.

2. High instantaneous powers are observed in the work loop experiments. In some cases researchers consider that those are comparable to the one obtained from animal movement.

3. Power outputs of tongue projection of reptile are extremely high as 9,600 W/kg and 18,129 W/kg. There must be a mechanism of elastic storage in the system.

4. Many factors are involved in transforming power from muscle fibre level to animal movement level. No general theory has not yet been constructed that explain power output of animal's movement from the power of muscle.

Chapter 6   Mechanical Power Output of Muscle

References

1. Marsh, R. L. and John-Alder, H. B., Jumping performance of hylid frogs measured with high-speed cine film, J. Exp. Biol. (1994) 188: 131-141.

2. Peplowski, M. M. and Marsh, R. L. Work and power output in the hindlimb muscles of cuban tree frogs Oseopilus septentrionalis during jumping. J. Exp. Biol. (1997) 200: 2861–2870.

3. James, R. S. and Wilson, R. S. Explosive Jumping: Extreme Morphological and Physiological Specializations of Australian Rocket Frogs (Litoria nasuta). Physiol. Biochem. Zool. (2008) 81: 176-185.

4. Askew, G. N., March, R.L., and Ellington C, P., The mechanical power output of the flight muscles of blue-breasted quail (Coturnix chinensis) during take-off. J. Exp. Biol. (2001) 204: 3601-3619.

5. Roberts, T. J. and Scales, J. A., Mechanical power output during running accelerations in wild turkeys., J. Exp. Biol. (2002) 205:1485-94.

6. Henry, H. T., Ellerby, D. J., and Marsh, R. L., Performance of guinea fowl Numida meleagris during jumping requires storage and release of elastic energy, J.Exp.Biol. (2005) 208:3293-302.

7. Curtin, N. A., Woledge, R. C., and Aerts, P., Muscle directly meets the vast power demands in agile lizards, Proc. Biol. Sci., (2005) 272(1563): 581–584.

8. McGowan, C. P., Baudinette, R. V., Usherwood, J. R., and , Biewener AA. The mechanics of jumping versus steady hopping in yellow-footed rock wallabies.   J. Exp. Biol. (2005) 208: 2741-51.

9. Lappin, A. K., Monroy, J. A., Pilarski, J. Q., Zepnewski, E. D., Pierotti, D. J., and Nishikawa, K. C. Storage and recovery of elastic potential energy powers ballistic prey capture in toads. J. Exp. Biol., (2006) 209: 2535-2553.

10. Stephen M. Deban, S. M, O'Reilly, J.C., Dicke, U., and Leeuwen, J. L., Extremely high-power tongue projection in plethodontid salamanders, J. Exp. Biol. (2007) 210: 655-667.

11. Morris, C. R. and Askew, G. N., The mechanical power output of the pectoralis muscle of cockatiel (Nymphicus hollandicus): the *in vivo* muscle length trajectory and activity patterns and their implications for power modulation. J. Exp. Biol. (2010) 213: 2770-2780.

12. Askew, G. N. and Marsh, R.L., The mechanical power output of the pectoralis muscle of blue-breated quail (Coturnix chinensis): the *in vivo* length cycle and its implications for muscle performance. J. Exp. Biol. (2001) 204: 3587-3600.

Chapter 6 Mechanical Power Output of Muscle

13 Jackson, B. E. and Dial, K. P., Scaling of mechanical power output during burst escape flight in the Corvidae, J. Exp. Biol. (2011) 214: 452-461.

14 Ellerby, D. and Askew, G. N., Modulation of flight muscle power output in budgerigars Melopsittacus undulates and zebra finches Taeniopygia guttata: in vitro muscle performance, J. Exp. Biol. (2007) 210: 3780-3788.

15 Hill, A.V., The heat of shortening and dynamics constants of muscles. Proc. R. Soc. Lond. B. (1938) 126; 136–195.

16 Josephson, R., K., 1985, Mechanical power output from striated muscle during cyclic contraction, Journal of Experimental Biology 114 (1985) 114: 493-512.

17 Machin, K.E. and Pringle, J., W., S., The physiology of insect fibrillar muscle. III. The effect of sinusoidal changes of length on a beetle flight muscle. Proc. R. Soc. Lond.B (1960) 152: 311–330.

18 Jewell, B.R. and Rüegg, J. C., Oscillatory contraction of insect fibrillar muscle after glycerol extraction. Proc. R. Soc. Lond.B, (1966) 164: 428–459.

19 Steiger, G. J. and Rüegg J. C., Energetics and "efficiency" in the isolated contractile machinery of an insect fibrillar muscle at various frequencies of oscillation. Pflugers Arch., (1969) 307(1):1-21.

20 Kawai, M. and Brandt, P. W., Sinusoidal analysis: a high resolution method for correlating biochemical reactions with physiological processes in activated skeletal muscles of rabbit, frog and crayfish. J. Muscle Res. Cell Motility (1980) 1: 279–303.

21 Askew, G. N. and Marsh, R. L., The effects of length trajectory on the mechanical power output of mouse skeletal muscles, J. Exp. Biol. (1997) 200:3119-31.

Chapter 6 Mechanical Power Output of Muscle

Chapter 7   Muscle – Tendon Complex

Recent advances of experimental devices enable real-time measurement of muscle and related tissues *in vivo* or *in situ*. Especially, it has been revealed that the function of tendon plays crucially important role in the movement.   In the previous chapter, it is implied that elastic storage should be involved, because the mechanical power measurement of movement of the individual seems to excess available muscle power output. However, its mechanism, i.e. what element actually works in movement of the individual is not fully investigated. From 1990's developments of real-time measurements *in vivo* became to reveal such mechanism. In this chapter, newest experimental results obtained *in vivo* is introduced.

7.1   What role tendon plays while muscle is shortening ?

The author is familiar with physics, then, it is a little curious to hear a terminology "isometric force" which is widely used in muscle physiology. The muscle produces force by shortening, but the isometric force is measured in the condition of constant length. Then, what mechanism is involved in both condition of shortening and constant length ?

Several kind of tissues may be involved in the process, but, probably the most important factor in it should be tendon. Consider the case that a tendon is serially connected to a muscle as schematically shown in Fig.7.1.   When the muscle occur voluntary contraction, then the tendon lengthens in the condition of constant length. This is the mechanism that the muscle produces force in the isometric shortening. (The author is informed of this by the publication of Prof. Tetsuo Fukunaga, who is the president of Kaya sports Univ in Japan at present.)

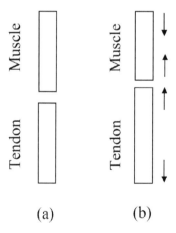

Fig.7.1 A mechanism of isometric contraction. When a muscle shortens, a tendon lengthens. Then, the force produced in the is condition of constant length.

Chapter 7 Muscle – Tendon Complex

This idea was originally implied by Hill in 1959 as visco-elastic property of muscle system (Hill, 1959), however, the real-time direct measurements *in vivo* have been realized from 1990's.

In 2001 Fukunaga et al. achieved real-time measurements of length changes in the fascicle and tendon of the human gastrocnemius medialis muscle during walking (Fukunaga et al. 2001). This experiment *in vivo* was realized by using real-time ultrasonography. The system is composed of muscle, free tendon and aponeurosis in both distal and proximal ends, which is later called as MTC (muscle-tendon complex). In the first part of this chapter, their group's result obtained in the movement of human jumping is explained.

Fig.7.2 represents time histories of muscle-tendon complex (MTC) during human vertical jumping (Kurosawa et al., 2001). The data are (a) MTC length, (b) fascicle length, (c) tendinous structure length, (d) fascicle angle, (e) velocity, (f) force and (g) mechanical power, respectively. The vertical bar in (a)−(d) represents SE for n=8 subjects.

Time is expressed relative to the instant of toe-off.

Time=0 　　　　　　　Toe-off of jumping.
Time= - 100 (msec)　　Onset from which the MTC length (a) remarkably shortens.
Time= - 350 (msec)　　Onset of starting upward movement.

At Time=−50 (msec). the mechanical power of MTC (muscle−tendon complex) denoted as (■) reaches the maximum in Fig.7.2(g). It is noticed that the tendinous structures denoted as (▲) contributes to the maximum in Fig.7.2(g). On the contrary, muscle fascicles power denoted as (●) does not contribute to this peak. Rather, the peak of muscle fascicles power (●) locates significantly in advance to the peak of MTC (■), which is around Time =−130 (msec). Interestingly, power of tendinous structures shows negative peak at this time. The positive peak of muscle fascicles and negative peak of teninous structures make cancellation of mechanical power. It means that if we calculate mechanical power from any of a mechanical model constructed by limb segments, it does not predict activity of muscle fascicles. Because, such a model would predict that mechanical power around Time =−130 (msec) is nearly zero. This finding is crucially important on the mechanism of muscle and tendon structure in the movement of individual. Kurokawa et al. summarized on the functions of each element as follows;

Chapter 7 Muscle – Tendon Complex

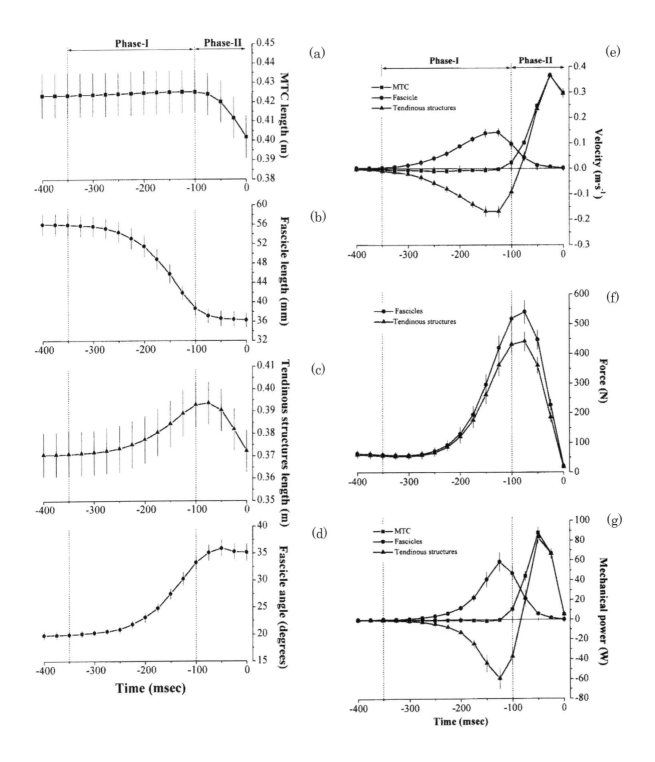

Fig.7.2 The data obtained by real-time measurement *in vivo* during the movement of human jumping (Kurosawa et al., 2001). (a) MTC (muscle−tendon complex), (b) fascicle length of muscle, (c) tendinous structures length, (d) fascicle angle, (e) velocity, (f) force and (g) mechanical power. In (e)−(g) each symbol represents ■ MTC (muscle−tendon complex), ● Fascicles and ▲ Tendinous structures.

Chapter 7　Muscle – Tendon Complex

1. At the time of maximum mechanical power of MTC (time= - 50 msec), muscle fascicles shorten isometrically, namely, constant length (fig.7.2(b)) and large force (Fig.7.2(f)).

2. At time=−50 msec, the tendon structure length changes rapidly (Fig.7.2(c)), and then, MTC length changes greatly (Fig.7.2(a)). This brings large velocity of MTC (Fig.7.2(e)) and consequently, brings the peak of mechanical power (Fig.7.2(g)).

    Muscle fascicle has a property of hyperbolic relation of force and velocity as in Hill's equation. Then, muscle can produce a large force for slow speed contraction. The time when MTC generates the maximum power is the timing that muscle produces large force in the condition of slow speed.

3. At time= - 150 msec, in advance to the time when the MTC reaches the maximum power, the muscle fascicles shorten (Fig.7.2(b)). At this time, the muscle fascicles produce large mechanical power (Fig.7.2(g)).

4. When the muscle fascicles produce the maximum power (time=-150 msec), MTC power is nearly zero (Fig.7.2(g)).

Then, a mathematical model involving only the function of muscle fails to predict individual power output in movement. Including the passive element such as tendon is required to understand the power output of individual in movement.

7.2　What happen to muscle and tendon during walking?
−The tendon supplies joint power during stance phase of walking−

From 1980's modern experimental devices enable real-time measurements of various properties of muscles and surrounding tissues *in vivo*. The following devices make possible to observe each quantity followed by arrow,

- Ultrasonography
    - → muscle fascicular length
    - →　tendon length

Chapter 7   Muscle – Tendon Complex

  → (pennation angle, and muscle thickness)
- EMG (electromyographical) activity
  → muscle activity
- Electrogoniometer
  → Ankle joint and knee joint rotation
- Force plate
  → GRF (ground reaction force)

In 2001 Fukunaga et al. published measurements data *in vivo* during human walking. As a result, it has been revealed that the tendon plays an important role in the movement of walking (Fukunaga et al., 2001). Fig.7.3 displays the published result, and it tells how muscles and surrounding tissues collaboratively work in the movement of individual.

In Fig.7.3(a) thick line shows GM (gastrocnemius medialis) fascicular length change in time. The dashed line and thin line show MTC and tendon length changes, respectively. Fig.7.3(b), (c) and (d) show EMG (electromyography) activity, joint angle and ground reaction force during walking, respectively. In Fig.7.3(b) EMG activity is prominent in the period of time=0.6~1.0 (s), which means muscle activity is high in this period. The bold line (GM fascicular length) in Fig.7.3(a) shows relatively flat level, which means muscle is in the state of isometric contraction. In this period velocity of muscle shortening is low, and the force of muscle is large. This implies that power output of the muscle is low, because power is defined by time derivative of the work.

On the contrary, the dashed line (MTC) and the thin line (tendon) of Fig.7.3(a) show the increase during time=0.4~1.0 (s). The behaviors of these two show similar tendency. This reveals that the length change of MTC is attributed to the length change of tendon. Then, it is concluded that joint power during stance phase of walking is supplied by power of tendon, i.e. lengthening movement of tendon. We summarize for the function of the muscle and the tendon during walking as follows. Note that the muscle below means GM (gastrocnemius medialis), and the tendon means free tendon and aponeurosis in both distal and proximal ends.

1   Joint power during stance phase in walking is mostly supplied by the power of the tendon lengthening.

2   During this time, the muscle does not generate power greatly. The muscle is in the state of isometric contraction (being in constant length), and it results in producing large force.

Chapter 7 Muscle – Tendon Complex

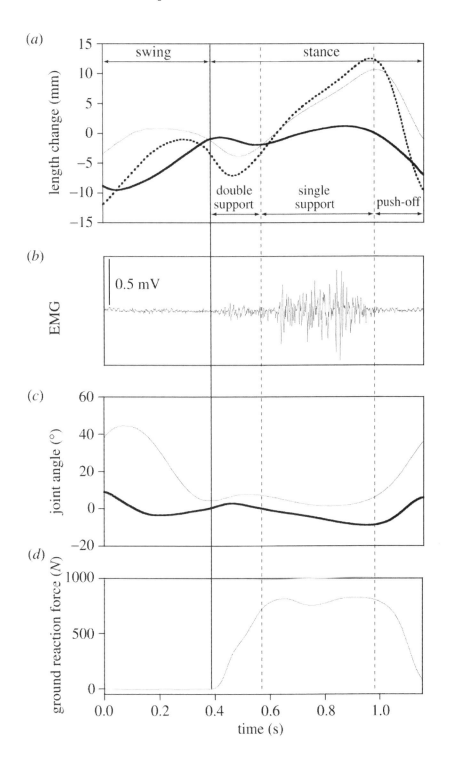

Fig.7.3 Each data obtained *in vivo* for the stance phase of human walking (Fukunaga et al., 2001). Each figure shows the following; (a) length change, (b) EMG (electromyography) activity, (c) joint angle, (d) ground reaction force. Each line in (a) shows the following; bold line – GM fascicular length, thin line – tendon length, and dashed line – MTC (muscle-tendon complex) length. In (c) thick line and thin line show ankle and knee joint angle, respectively.

Chapter 7 Muscle – Tendon Complex

References

Fukunaga, T., Kubo, K., Kawakami, Y., Fukashiro, S., Kanehisa, H., and Maganaris, C.N. 2001. In vivo behaviour of human muscle tendon during walking, Proc. R. Soc. Lond. B (2001) 268: 229-233.

Kurosawa, S., Fukunaga, T., and Fukashiro, S., Behavior of fascicles and tendinous structures of human gastrocnemius during vertical jumping. Journal of Applied Physiology, (2001) 90: 1349-1358.

Kurosawa, S., Fukunaga, T., Nagano, A., and Fukashiro, Interaction between fascicles and tendinous structures during counter movement jumping investigated in vivo, J. Appl. Physiol., (2003) 95: 2306-2314.

Chapter 7   Muscle – Tendon Complex

Chapter 8 Evaluation of running ability for *T.rex*

Chapter 8 Evaluation of running ability for *T.rex*

In this chapter and the following chapters, the author presents the assessment of locomotion ability of *Tyrannosaurus* based on the knowledge described in the previous chapters 1~7. *Tyrannosaurus* is a name of the genus, but only one species is known in the genus. *Tyrannosaurus rex* is a name for the species. *Tyrannosaurus rex* is called as *T.rex* for short, then, we call *T.rex* which is the target to evaluate running ability in this book.

8.1 Estimation of running speed of *T.rex* in the literature

At first, let us see how it is described on running speed of *T.rex* in the published scientific articles. In 1986 and 1988 Bakker and Paul proposed 45 mph and 40 mph running of *T.rex*, respectively, which correspond to the speed of 20 m/s and 17.9 ($\cong$ 18) m/s (Bakker 1986, Paul 1988). This estimation is quite fast because human's fastest speed is around 10m/s. (Note that Usain Bolt's top speed was 12.27 m/s achieved in 2009.) Their arguments were based on the morphological consideration of muscle and limb structure. Hence, reliability of speed estimation is limited. Then, such a question may be raised that, "Are there any fossil evidences which show running motion of dinosaurs ?"

Unfortunately, evidences of running dinosaur are quite limited. It is because that running motion is rather unusual motion for wild animal. Most of the foot prints show 1~3m stride length of dinosaurs, which shows the result of walking motion. In the published literature two evidences of running motion of theropod (bipedal carnivorous dinosaur) have been reported, which were marked in Mesozoic strata. Farlow reported that 6.59 m stride length trackway of theropod was observed in the formation of lower cretaceous at Texas, U.S. (Farlow 1981). The average foot length was 38 cm, which yielded the velocity of 11.1 m/s for locomotion speed of the theropod. Another foot print with wide stride length was reported by Day et al. (Day et al., 2002). 5.65 m stride length made by theropod was observed in the formation of middle Jurassic at Oxfordshire, UK. The estimated speed was 8.11 m/s by the use of Alexander relation which will be explained in the next subsection.

1 Bakker proposed 20 m/s, and Paul proposed 18 m/s as a speed estimation of *T.rex* from the morphological consideration of bones.

2 11.1 m/s and 8.11 m/s running evidences have been found in the fossilized footprint of medium or large size theropod. (Alexander relation was used for speed estimation.)

Chapter 8　Evaluation of running ability for *T.rex*

8.2　Alexander's qualitative speed evaluation method
　　— An introduction of Froude number—

Alexander established a monumental theory on dinosaur locomotion. In the theory, a quantity called Froude number $Fr = \dfrac{V^2}{g \cdot h}$ plays an important role, where V, h, and g are velocity, characteristic length of a leg and gravitational constant, respectively. This is an dimensionless parameter, i.e., $Fr = \dfrac{V^2 \,[\mathrm{m^2/s^2}]}{g\,[\mathrm{m/s^2}] \cdot h\,[\mathrm{m}]}$ which characterize dynamical motion. (Another famous dimensionless parameter is Reynolds number $Re = \dfrac{V\,[\mathrm{m/s}] \cdot h\,[\mathrm{m}]}{\nu\,[\mathrm{m^2/s}]}$ which is used in fluid dynamics, where ν represents kinematic viscosity.)

He employed dynamic similarity hypothesis such that the animal walked or ran in a dynamically similar fashion at the same Froude number (Alexander, 1976; Alexander et al., 1983). With comparing extant animal data, he found that relative stride L_{st}/h has a relation with Froude number as,

$$\frac{L_{st}}{h} = 2.3 \left(\frac{v^2}{g \cdot h}\right)^{0.3} , \quad (8.1)$$

where L_{st} is the stride length which is the distance between two successive falls of the same foot. The characteristic length h is usually set as hip height. In the recent data of various birds and human the exponent seems to be close to 0.5 rather than 0.3 (Gatesy and Biewencr, 1991). This situation leads that the relative stride length L_{st}/h has a linear relation with relative velocity as,

$$\frac{L_{st}}{h} \propto \left(\frac{v^2}{g \cdot h}\right)^{0.5} . \quad (8.2)$$

However, following researches have revealed that several factors may cause an error in estimating locomotion speed from stride length using this relation. For the improvement, many methods and discussions have been presented to the first version of Alexander's one (Gatesy and Biewener, 1991; Russell and Beland, 1976; Thulborn, 1981; Thulborn, 1989; Rainforth and Manzella, 2007; Wallace and Brooks, 2003).

Chapter 8 Evaluation of running ability for *T.rex*

In recent review Alexander stated as follows:

"The method cannot claim to be accurate." "Unfortunately, the other potential sources of error remain serious. Though it cannot predict precise speeds, the method is informative; there seems to be no likelihood of confusing a stroll with a sprint" (Alexander, 2006).

This point will be discussed in Sec.11.4 in detail. The summarize of this subsection is as follows;

1. Froude number is defined as $Fr = \dfrac{v^2}{g \cdot h}$.

2. Some relation such as $\dfrac{L_{st}}{h} \propto \left(\dfrac{v^2}{g \cdot h}\right)^{0.5}$ has been proposed by Alexander et al.

In the above expressions, L_{st} and h represent the stride length between two successive falls of the same foot and the hip height, respectively.

8.3 How we estimate hip height *h* from foot print?

On the reconstruction of locomotion of bipedal dinosaur from footprint, several researchers have thrown a warning to use Alexander relation (Alexander, 1976) for quantitative speed estimation. To use this relation, hip height h must be estimated from foot length FL.

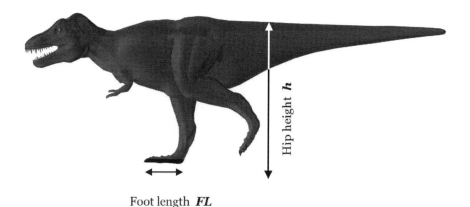

Foot length **FL**

Fig.8.1 The hip height h and the foot length FL. Alexander proposed a relation of $h = 4FL$ (Alexander, 1976).

Alexander's first proposal was $h=4FL$, namely, hip height is four times foot length as shown in Fig.8.1 (Alexander, 1976). However, Thulborn stated that the factor ranges from 4.5 to 5.9 according to type and size of the dinosaur (Thulborn, 1989). Recently, Rainforth and Manzella re-analyzed this factor using 24 specimens from different dinosaurian groups, and concluded that speed estimation could be incorrect by a factor of two (Rainforth and Manzella 2007).

They stated that "there is no reliable way to estimate hip height from footprint length, either using morphometric or allometric equations (Rainforth and Manzella 2007)". Thus, it should be kept in mind that speed estimation using Alexander relation and Froude number may involve large uncertainty. As a summary, the followings are stated.

1. The hip height h is estimated by the relation $h=4FL$, where FL is the foot length, which was originally proposed by Alexander.

2. However, following studies revealed that there is a large uncertainty in this relation up to a factor of 2.

8-4 Computer simulation studies

In 2007 Sellers et al. achieved dynamical calculation, and found that *T.rex* with 6 ton mass would be possible to run with speed of 8m/s～10m/s (Sellers et al., 2007). The speed range depends on an assumption how much muscle mass *T.rex* had on limbs. Sellers et al's work was the first study to calculate whole running motion of *T.rex* based on biomechanical theory. However, unfortunately, parameters that they used were not described satisfactorily, it is difficult to reconstruct their result for following researchers. Then, the author discusses validation of parameters involved in the running simulation in Chap.10~13 in detail.

A summary of published results is shown in Fig.8.2. In this graph, horizontal axis and vertical axis represent running speed and Froude number, respectively. I add twofold error bar for Farlow and Day et al.'s observation data, which are expressed as the horizontal line. I draw shadow which represents the range of human running speed. From this figure, we notice that if *T.rex* and human run a race, *T.rex*'s speed might be comparatively similar or even faster than human's speed. Then, such a person would feel that *T.rex* run fast. However, Hutchinson and collaborators cast a doubt for fast running ability of *T.rex* as described in the next chapter.

Chapter 8 Evaluation of running ability for *T.rex*

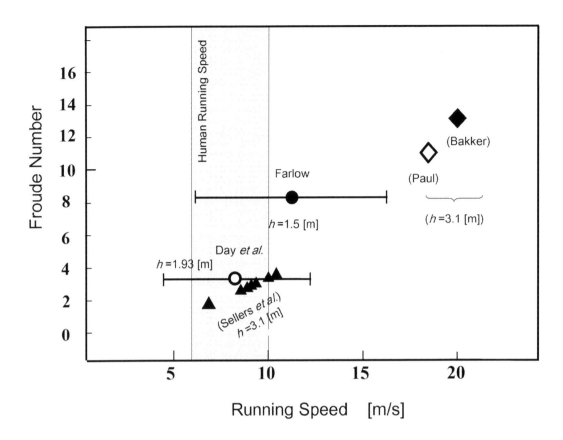

Fig. 8.2. Published estimation of large and medium size *T.rex* running speed. The vertical axis represents Froude number, $Fr = \dfrac{v^2}{g \cdot h}$, where v, g and h are velocity, gravity constant and hip height, respectively. Human running speed is in the range 6 m/s ~10 m/s. Hence, *T.rex* running speed is similar or faster than the human's one in these estimations.

In concluding this chapter, the references are listed which relate to bipedal and quadrupedal dinosaurs locomotion ability (Alexander, 1983, 1985, 1989, 1996; Gatesy, 1991; Farlow et al., 1995; Farlow et al., 2000; Christiansen, 2000; Gatesy and Biewener, 1991; Russell and Beland, 1976; Wallace and Brooks, 2003).

References

Alexander,R. Mc. N. 1976. Estimates of speeds of dinosaurs. Nature 261: 129-130.

Alexander, R. Mc. N. & Jayes, A. S. A dynamic similarity hypothesis for the gaits of quadrupedal mammals, *J. Zool.* 201, 135-152(1983).

Alexander, R. Mc. N. Mechanics of posture and gait of some large dinosaurs, *Zool. J. Linn. Soc.* 83: 1-25(1985).

Chapter 8 Evaluation of running ability for *T.rex*

Alexander, R. Mc. N. *The Dynamics of Dinosaurs and Other Extinct Giants* (Columbia University Press, New York, 1989).

Alexander, R. Mc. N. Tyrannosaurus on the run. Nature, (1996) 379: 121-121.

Alexander, R. Mc. N. Dinosaur biomechanics, *Proc. Roy. Soc.,* (2006) B 273: 1849-1855.

Bakker, R. T. *Dinosaur Heresies* (William Morrow, New York, 1986).

Christiansen, P. Strength indicator values of theropod long bones, with comments on limb proportions and cursorial potential, *Gaia,* (2000) 15: 241-255.

Day, J. J., Norman, D. B., Upchurch, P. & Powell, H. P. Dinosaur locomotion from a new trackway, *Nature,* (2002) 415: 494-495.

Farlow, J. O., Smith, M. B. & Robinson, J. M. Body mass, bone "strength indicator", and cursorial potential of Tyrannosaurus rex, *J. Vert. Paleo.*, (1995) 15: 713-725.

Farlow, J. O. Estimates of dinosaur speeds from a new trackway site in Texas. *Nature,* (1981) 294: 747-748.

Farlow, J. O., Gatesy, S. M., Holtz, T. R. Jr., Hutchinson, J. R. and Robinson, J. M. 2000. Theropod locomotion, *Am. Zool.*, (2000) 40: 640-663.

Gatesy, S. M. Hind limb scaling in birds and other theropods: Implications for terrestrial locomotion, *J. Morph.*, (1991) 209: 83-96.

Gatesy, S. M. & Biewener, A. A. Bipedal locomotion: effects of speed, size and limb posture in birds and humans, *J. Zool., Lond.,* (1991) 224: 127-147.

Paul, G. S. *Predatory Dinosaurs of the World* (Simon & Schuster, New York, 1988).

Paul, G. S. Limb design, function and running performance in ostrich-mimics and tyrannosaurs, *Gaia,* (2000) 15: 257-270.

Rainforth, E. C. & Manzella, M. Estimating speeds of dinosaurs from trackways: a re-evaluation of assumptions. in (Rainforth E. C. ed.) Contributions to the paleontology of New Jersey (II), pp41-48 (GANJ 24, 2007).

Russell, D. A. & Beland, P. Running Dinosaurs, *Nature,* (1976) 264: 486.

Thulborn, R. A. Estimated speed of a giant bipedal dinosaur, *Nature* 292, 273-274(1981).

Thulborn, R. A. 1989. The Gais of dinosaurs. in *Dinosaur Tracks and Traces* (Gillette, D. D. & Lockley ed.) (Cambridge University Press, 1989).

Thulborn, R. A. *Dinosaur Tracks* (Chapman & Hall, London, 1990).

Wallace, R. L. & Brooks, W. S. A dinosaur trackways exercise, *Bioscene* 29, 3-7(2003).

Sellers, W. I. and Manning, P. L., Lyson, T., Stevens, K. & Margetts, L. Virtual Palaeontology: Gait Reconstruction of Extinct Vertebrates Using Hight Performance Computing, *Palaeontologia Electronica* 12.3.13A: 1-14(2009).

Chapter 9 Casting a doubt for fast running ability of *T.rex*

9.1 Was *T.rex* not a fast runner?

In 2002 Hutchinson et al. presented the first quantitative study on running ability of *T.rex* (Hutchinson and Garcia, 2002). They applied static calculation for the posture of running motion of *T.rex*. In the study they assumed an arbitrary mid-stance posture in running motion, and calculated muscle mass to support the posture. As a result, they concluded that a fraction of 43 % of body mass is needed for the muscle mass of one leg in order to run quickly.

Recent mass distribution studies reported the following values for one leg relative to the whole mass; 16.0 % for MOR555 (Bates et al. 2009), 14.2 % for MOR 555 (Hutchinson et al. 2007), 14.4 % for BHI3033 (Bates et al. 2009). The abbreviation stands for Museum of Rockies and Black Hills Institute, respectively. Then, according to their argument *T.rex* had not enough amount of muscle for quick running, then, *T.rex* would was not a fast runner. The estimated running speed which *T.rex* could not achieve is 20 m/s with the use of an discussion of Froude number. With a reference of ostrich motion as $Fr=16$, *Tyrannosaurus* hip height $h=2.5$ leads to $v=20$ m/s as $16 \simeq 20^2 / 9.8 \cdot 2.5$. Then, *T.rex* could not run in a speed of 20 m/s.

The issue was studied again in 2004 based on the same methodology (Hutchinson, 2004a, 2004b). In the study Hutchinson corrected required muscle mass as 21 % for one leg. It can be said that there is 49 % ($=21/43 \times 100$) uncertainty in predicting required muscle mass in the evolution of the study. In 2009, Gatesy et al. calculated the value for various mid-stance postures (Gatesy, Baker and Hutchinson, 2009). The methodology is the same with Hutchinson and Garcia's, then, the speed estimation is heuristic. The results are considered as medium speed running is possible, because 18.3 ($\sim 1.87 \times 9.8$) m/s² vertical acceleration is allowed in their estimation for static postures.

For validity of this theory the following questionable points are raised;

1 Their theory is static one. Then, it can not bring any information on speed estimation in principle.

   (No explicit speed variable v appeared in the main framework of static calculation.)

2   Speed evaluation is achieved by the use of simple Froude number discussion, $16 \cong 20^2 / 9.8 \cdot 2.5$. This is qualitative evaluation, not quantitative. And, it should be remind that such evaluation contains error in a certain range.

3   Scientific data usually contains error. The data is expressed as mean value ± standard deviation (SD) or standard error of the mean (SEM). On the contrary, no one has ever evaluated an error in this evaluation.

4   It is reminded that the value was corrected as 21% in the article (Hutchinson, 2004b). It is only two years later from the first publication in which the value was published as 43%.

9.2   Basic biomechanics for the static theory

In 2002 Hutchinson and Garcia published a scientific paper entitled "*Tyrannosaurus* was not a fast runner" (Hutchinson and Garcia, 2002).   In this subsection the author explains the content of the theory.

Hutchinson et al's estimation for running ability of *T.rex* is based on static method (Hutchinson and Garcia, 2002; Hutchinson, 2004b; Gatesy et al. 2009).   They postulate a mid-stance posture arbitrarily among many possibilities, and calculate required muscle mass to support the posture. A model of *T.rex* is made by trunk, thigh, shank, metatarsus and foot, as shown in Fig. 9-1(a). Each segment length is 1.13, 1.26, 0.699 and 0.584 m for thigh, shank, metatarsus and foot, respectively. The angle between each segment is chosen as 50, 110, 140, 80 (degree) for hip, knee, ankle and toe, respectively, for best guess model (Hutchinson and Garcia, 2002; Hutchinson, 2004b).

For calculating joint torque, or moment of the force about joint, a free-body diagram analysis is applied as shown in Fig.9.1(b). For example, let call the foot segment as segment "1", and define the mass and the moment of inertia as m_1 and I_1, respectively. Then, the equations of motion for translation and rotation become as follows in (x, y) plane,

$$\vec{F}_1 - \vec{F}_2 - m_1 g \big|_y = m_1 \vec{a}_1 \qquad \text{(9.1a)}$$

$$\vec{x}_{1g} \times \vec{F}_1 - \vec{x}_{2g} \times \vec{F}_2 + M_1 - M_2 = I_1 \dot{\omega}_1 \qquad \text{(9.1b)}$$

Chapter 9 Casting a doubt for fast running ability of T.rex

where $\vec{F}_1$, $\vec{F}_2$ and $\vec{a}_1$ are the force from downside segment, the force from upper segment and acceleration, respectively.

For rotational motion, $\vec{x}_{1g}$ and $\vec{x}_{2g}$ are the vector from the center of mass of 1-th segment to the point acting force $\vec{F}_1$ and $\vec{F}_2$. M_1 and M_2 are the moment of force between 0-th and 1-th, 1-th and 2nd segment, respectively. For the case of 1-st segment $\vec{F}_1$ corresponds to the ground reaction force. $\dot{\omega}_1$ is time derivative of angular velocity. For usual running motion right side term $m_1\vec{a}_1$ and $I_1\dot{\omega}_1$ are small compared to left side term. Then, letting $m_1\vec{a}_1 = 0$ and $I_1\dot{\omega}_1 = 0$ is fairly justified. Thus, putting known terms $\vec{F}_1$ and M_1 into Eq.(9.1) yields unknown terms $\vec{F}_2$ and M_2. Iterating calculation yields the moment of force M_i for i-th segment, subsequently.

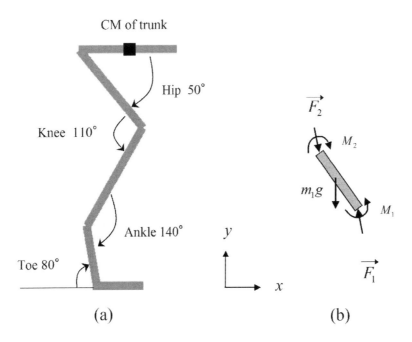

Fig. 9.1. Hutchinson et al's segment model of T.rex leg (a), and a free-body diagram (b). (a) is a best guess of mid-stance posture in running motion in the reference (Hutchinson and Garcia, 2002). Using (b) the torque at a joint can be calculated from foot segment, sequentially.

The ground reaction force is assumed to be 2.5 times whole mass times gravity for 20 m/s running with a reference to ostrich running. Then, a fraction of muscle mass for i-th joint to the total body mass m_i (%) is calculated as the following equation (Hutchinson and Garcia, 2002),

Chapter 9   Casting a doubt for fast running ability of *T.rex*

$$m_i(\%) = \frac{100 M_i L d}{\sigma c r m_{body} \cos\theta} \quad , \quad (9.2)$$

An introduction of this expression is explained in detail in Appendix of Chap.4. In the expression, relatively reliable factors for the final result are the following two. The first is muscle density $d=1.06\times 10^3 \text{kgm}^{-3}$, and the second is the fraction of active muscle volume c. A concept of estimating the minimum muscle leads the setting of $c=1$. The total body mass m_{body} is not a intrinsic parameter, because the joint moment M_i includes $m_{body}\times$(gravitational acceleration) term, which leads cancellation of the factors. Although, the expression does not contain total mass factor, the theory is intended to apply for *Tyrannosaurus* (*T.rex*) having total mass of 6 ton. It is noted that Bates et al's mass distribution analysis showed that *T.rex* (MOR555) would have 16 % mass relative to the total mass for one leg (Bates, et al., 2009). Then, only *T.rex* whose mass is smaller than 2600 kg could run, which is shown as the gray area in Fig.15.1. On the other hand, body weight heavier than 2600 kg *T.rex* could not run.

Note that the following values of $m(\%)$ are published: 14.2 % for MOR 555 (Hutchinson et al., 2007) , 14.4 % for BHI3033 (Bates et al. 2009), and 14.6~25.4 for Carnegie, Sue, Stan and MOR (Hutchinson et al., 2011).

9.3   Uncertainty of the parameters *L*: muscle fibre length

As the author noted in Sec.9.1, scientific data usually contains error. The data is expressed as mean value ± standard deviation (SD) or standard error of the mean (SEM). On the contrary, no one has ever evaluated an error in this evaluation. In the subsequent subjections the author calculate standard deviation of Hutchinson et al's work.

The *L* is muscle fibre length (Hutchinson and Garcia, 2002), which was re-defined as muscle fascicle length, later (Hutchinson, 2004b). As described in Chap.1, muscle fascicle is equal to bundle of fibres, and it is composed of muscle fibres. However, the length of muscle fascicle is usually called "fibre length" in the field of muscle physiology which relates to biomechanics described in Chap.2~7. So, those are the same meaning, usually.

More important problem lies in another point. At present, we can have clear 3D image of muscle by MRI. It is flexible object having complex boundary, and the concept of the fibre length and the cross-sectional area is an averaged quantity over the captured 3D image of the muscle. Then, using muscle volume which a quantity directly measured to evaluate locomotion ability is said to be more appropriate than using averaged quantity muscle fibre length, which is described in detail in chap.5.

About muscle fascicle length *L*, Hutchinson listed the ratio of fascicle length/segment length

Chapter 9 Casting a doubt for fast running ability of T.rex

of 8 species of reptile, which is in the range of 0.367~1.13 (Hutchinson, 2004a). It is said that this variance is considerably large. This is due to the fact that morphology of those species greatly differs each other. The 8 species are *Basiliscus, Iguana, Alligator, Eudromia, Gallus, Meleagris, Dromaius and Stuthio*. We calculate standard deviation of L, and summarize its influence to the result in Table 9.1. The L and its standard deviation are 0.85 ± 0.31 m, 0.40 ± 0.12 m, 0.26 ± 0.13 m, 0.18 ± 0.07 m for thigh, shank, metatarsus and foot, respectively. We calculate how this variance affect to the final value of m_i. Hutchinson obtained 9.7 %, 2.7 %, 8.3 %, 7.1 % as mean value of m_i for thigh, shank, metatarsus and foot, respectively (Hutchinson 2004b). We put these into Eq.(9.2) and obtain M_i. And, we calculate standard deviation of m_i. Then, considering standard deviation of L, each m_i value is expressed as 9.6 ± 3.5 %, 2.6 ± 0.8 %, 8.3 ± 4.2 % and 7.0 ± 2.7 %. A little difference comes from significant figures in the calculation.

	hip/thigh	knee/shank	ankle/meta	toe/foot	Total
M [$\times 10^4$ Nm]	7.2	2.5	6.6	4.7	
L [m]	0.85 ± 0.31	0.40 ± 0.12	0.26 ± 0.13	0.18 ± 0.07	
$\cos\theta$	1				
r [m]	0.37	0.22	0.12	0.070	
σ [$\times 10^5$ N/m^2]	3.0				
m_i [%]	9.6 ± 3.5	2.6 ± 0.8	8.3 ± 4.2	7.0 ± 2.7	21 ± 5
σ [$\times 10^5$ N/m^2]	3.9				
m_i [%]*	7.4 ± 2.6	2.0 ± 0.6	6.4 ± 3.1	5.4 ± 2.1	16 ± 4

Table 9.1 Summary of a ratio of muscle mass for *i*-th segment of a leg per whole mass m_i. The M, L, θ, r σ are joint moment, mean extensor muscle fascicle length, pennation angle of muscle fibre, mean extensor muscle moment arm and maximum muscle stress, respectively. The standard deviation of muscle fibre length L is calculated from 8 species of the reference (Holland 1975; Hutchinson 2004a, 2004b). From m_i of the reference (Hutchinson 2004b), M_i is calculated. Then, the average and standard deviation of m_i is obtained. The lower column shows the case for $\sigma = 3.9 \times 10^5$ N/m^2, which is known large value of the maximum tetanic tension P$_0$ (James and Wilson 2008; Lännergren 1987).

Chapter 9 Casting a doubt for fast running ability of T.rex

Hutchinson stated that if each muscle is over 7%, the animal would not run quickly. Then, it is said that existence of these variations leads to the possibility that *T.rex* could run in a speed of 20 m/s in Hutchinson's framework.

Table 9.1 also includes an effect of variance of the maximum muscle stress, which is shown in lower column of Table 9.1. In the previous chapter and Chap.3 we have introduced that two independent research groups reported the maximum titanic tension larger than $P_o=39$ N/cm². Using this value for the evaluation, the evaluation of m_i becomes smaller. Finally, required muscle mass per one leg becomes 16 ± 4 %, as shown at the bottom of right row in Table 9.1. This implies that *T.rex* could run in a speed of 20 m/s even in the framework of Hutchinson's group estimation.

In the calculation described in this section, the author does not employ any special hypothesis. The only thing done is including published experimental results which are widely accepted. Especially, calculation of standard deviation is indispensable for the estimation of locomotion ability. In the experimental reports there are always descriptions of standard deviation or standard error of the mean. In the future works related to this issue the description of the variance is desirable.

Figure 9.2 shows the results of Table 9.1 graphically. Bates et al.'s estimation for *m* is 16 % which is shown as dark gray area in Fig.9.2, that was obtained by detailed *T.rex* mass distribution analysis (Bates et al., 2009). The result including known maximum of the tetanic tension Po and the standard deviation of L is shown at the top in Fig.9.2, which lies partly in the dark gray area. Then, we can not exclude possibility of 20 m/s running of *T.rex* based on Hutchinson's framework.

Chapter 9 Casting a doubt for fast running ability of *T.rex*

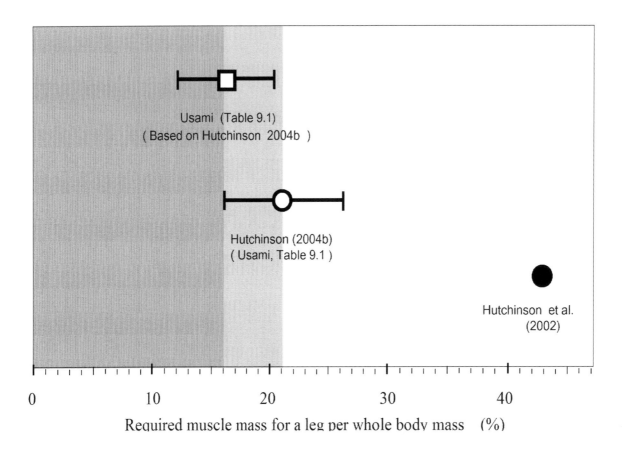

Fig.9.2 Re-evaluation based on Hutchinson's group analysis for required muscle mass for a leg per whole body mass. The gray areas represents Hutchinson's criteria 21 % that *T.rex* would have for one leg (Hutchinson, 2004b). The dark gray areas represents Bates et al.'s criteria 16 % that was obtained by detailed *T.rex* mass distribution analysis (Bates et al., 2009). The original estimation by Hutchinson and Garcia was 43%, shown as black circle at right hand side. Then, they concluded that *T-Rex* could not run fast. Two years later, One of the authors corrected the value down to almost half shown as white circle (Hutchinson, 2004b). The result including uncertainty of parameter L is shown at the middle in the graph. The data lies in the gray line, which tells that *T.rex* could run quickly in Hutchinson framework. Known maximum value of titanic tension P_0 is 3.9×10^5 N/m^2 which is described in Chap.3. The result including this value is shown at the top in the graph.

In 2011 Hutchinson et al's achieved 3D scanning of four adult and one juvenile specimens of well preserved *T.rex* skeleton, and analyzed their mass distributions (Hutchinson et al., 2011). Especially remarkable new calculation is an evaluation of amount of extensor muscle for a leg. Muscles are composed of extensor and flexor muscles. Then, the evaluation of extensor muscle

Chapter 9 Casting a doubt for fast running ability of *T.rex*

is a monumental result in this field. The result of ratio of extensor muscle mass relative to the whole mass is shown at the bottom in Fig.9.3. The upper three data in Fig.9.3 show theoretically required value of the ratio. In the work (Hutchinson et al., 2011), the most probable body mass estimations of four specimens (Carnegie, Sue, Stan MOR) are in a range of 6000 ~8000 ~ 9500 kg, which are heavier than the one that this book is assumed based on the early studies. A fraction of a leg muscle mass to the total body mass m (%) is in a range of 14.6~25.4, which is obtained by Hutchinson et al.'s work (Hutchinson et al., 2011). Then, an area less than m (%)=25.4 % is shown as gray in Fig.9.3. As seen from this graph, both theoretically required and the measurement data are overlapped. Then, it is not said that *T.rex* could not run fast.

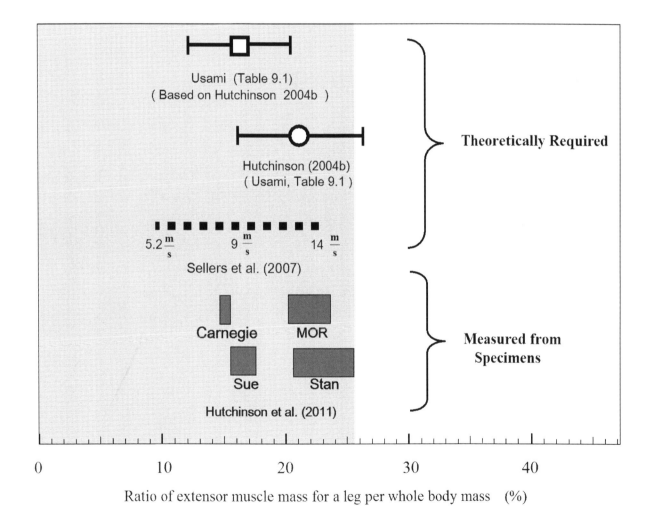

Fig.9.3 Ratio of extensor muscle mass for a leg per hole body mass. The lower four measured specimens are CM 9380 (Carnegie), FMNH PR 2081 (Sue), MOR 555 (MOR) and BHI 3033 (Stan) (Hutchinson et al., 2011).

Chapter 9 Casting a doubt for fast running ability of *T.rex*

9.4 Uncertainty of the parameters θ: pennation angle

Mean pennation angle of muscle fascicle θ is not well known parameter. Hutchinson et al. employed the value of 9.0, 21, 22, 21 (degree), for hip, knee, ankle and toe joint, respectively, with a reason of "measured from dissections for the extant taxa, or approximated for the extinct taxa" (Hutchinson and Garcia, 2002). However, zero value is used for the angles of all joint in the work with the following reason; "The term $\cos\theta$ is close to 1.0 in living animals, difficult to measure accurately ", "so it was left out ($\theta =0°$) as a simplifying conservative assumption " (Hutchinson, 2004b). Because, it is hard to fairly evaluate the value and the error of θ, we employ the value of $\theta=0°$ as Hutchinson et al.'s for the evaluation of m_i in our study.

9.5 Uncertainty of the parameters r: moment arm

The extensor muscle moment arm r changes its value as the change of joint angle. Hutchinson studied this subject in detail in the work of 2005 (Hutchinson et al. 2005). According to the work we summarized the data as Figs.9.4~9.6.

For hip extensor major 10 muscles are included, which are IT3, ILFB, FTI1, FTI3, FTE, ADD, ADD2, ISTR, CFB and CFL. Abbreviation is described below, for example, IT1 for Muscle iliotibialis 1. The mean value of those ten is shown as thick black curve in Fig.9.3, and standard deviation is shown as vertical line. If we represent the quantity of 10 muscles as one parameter, a summation of weighted value must be used. Then, the mean value shown in Fig.9.4~9.6 represents a general tendency of angle dependence of moment arm r.

Included muscles;

IT3	Muscle iliotibialis 3
ILFB	Muscle iliofibularis
FTI1,3	Muscle flexor tibialis internus 1,3
FTE	Muscle flexor tibialis externus
ADD1,2	Muscle adductor femoris 1, 2
ISTR	Muscle ichiotrochantericus
CFB	Muscle caudofemoralis brevis
CFL	Muscle caudofemoralis longus

Chapter 9 Casting a doubt for fast running ability of *T.rex*

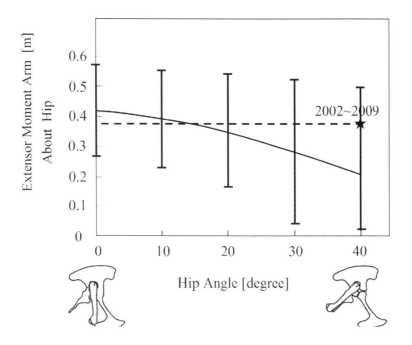

Fig.9.4 Angle dependence of an average of extensor moment arm *r* about hip joint. The 10 major muscles in the reference (Hutchinson et al. 2005) are included. The standard deviation is also shown as vertical solid line. The black star shows the data employed by Hutchinson et al's work (Hutchinson and Garcia 2002; Hutchinson 2004b; Hutchinson et al. 2005; Gatesy, Bäker, Hutchinson 2009).

As observed from the figure, *r* changes its value with the change of the angle. Hutchinson have published an average value of this in the works, namely, *r*=0.37m in the papers (Hutchinson and Garcia 2002; Hutchinson 2004b), and *r*=0.38m in the papers (Hutchinson et al. 2005; Gatesy et al. 2009). These are plotted as star in Fig.9.4. Hutchinson has used these average values for evaluation of running ability, however, Fig.9.4 shows discrepancy between solid curve and average value shown as star mark. Furthermore, standard deviation shown as vertical line is large. This uncertainty of the value of extensor muscle moment arm *r* yields uncertainty of the final evaluation of running ability of *T.rex*. In our running simulation presented in Chap. 10-13 average value of *r*=0.38m is employed for hip extensor muscle moment arm as the works of Hutchinson et al. Taking into account of the value represented as black curve leads to different estimation of running ability.

For knee extensor muscle moment arm the similar calculation is carried out, which is shown in Fig. 9.5. For this case 7 muscles are taken from Figure 5A of the reference (Hutchinson et al. 2005). In this case standard deviation is smaller than the hip case.

Chapter 9 Casting a doubt for fast running ability of *T.rex*

Included muscles;

IT1,2A,2P,3	Muscle iliotibialis 1-3
AMB	Muscle ambiens
FMTE	Muscle femorotibiales externus
FMTI	Muscle internus

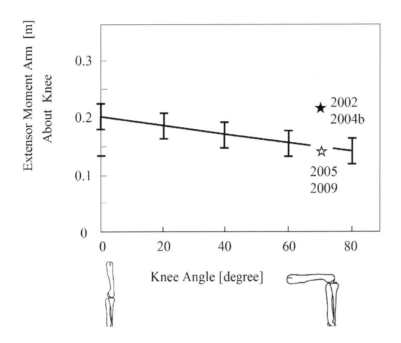

Fig.9.5 Angle dependence of an average of extensor moment arm *r* about knee joint. An average of 7 muscles is taken from Figure 5A of the reference (Hutchinson et al. 2005). The black star shows the data employed by Hutchinson et al's work (Hutchinson and Garcia, 2002; Hutchinson, 2004b). The white star shows the data employed by Hutchinson et al's work (Hutchinson et al., 2005; Gatesy, Bäker, and Hutchinson, 2009). The vertical line expresses standard deviation.

The employed value for running evaluation by Hutchinson et al.'s work has been changed. $r=0.22$m was used in the works (Hutchinson and Garcia, 2002; Hutchinson, 2004b) which is shown as black star in Fig.9.5, whereas $r=0.14$m was used in the works (Hutchinson et al., 2005; Gatesy et al., 2009). The discrepancy between them is remarkable as 57% (=(0.22-0.17)/0.17). Although, this discrepancy affected to the estimation of running ability of *T.rexs*, recent data show convergence to the similar values as $r=0.14$m. The solid line at the angle 70° is closely located around $r=0.14$m,

Chapter 9 Casting a doubt for fast running ability of *T.rex*

which is the mean value used in the reference (Hutchinson et al. 2005; Gatesy et al. 2009). Note that standard deviation is small for this case.

For ankle extensor moment arm similar tendency is observed in Fig.9.6 The employed value in the estimation of running ability by Hutchinson et al. has been changed from 2002's work to 2009's work, which are shown as white and black star in Fig.9.6. However, convergence of the value is also observed as white star mark and black curve. In our running simulation angle dependent value of r is employed for knee and ankle extensor moment arm, except for the calculation of Fig.12.1. For the case of Fig.12.1 average value of r in the work (Hutchinson 2004b) is employed for comparison with the result of work (Hutchinson 2004b). These parameters are summarized in Table 10.1

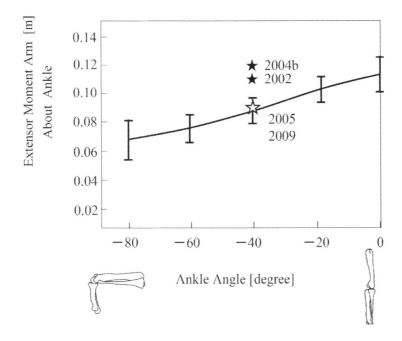

Fig.9.6 Angle dependence of an average of extensor moment arm r about ankle joint. An average of 6 muscles is taken from Figure 5C of the reference (Hutchinson et. al., 2005). The others are the same of Fig.9.4.

Chapter 9   Casting a doubt for fast running ability of *T.rex*

Included muscles;

    GL   Muscle gastrocnemius lateralis
    FL   Muscle fibularis longus
    FHL   Muscle flexor hallucis longus
    GM   Muscle gastrocnemius medialis

    FDL   Muscle flexor digitorum longus

References

Bates, K. T. Manning, P. L., Hodgetts, D., and Sellers, W. I., Estimating Mass Properties of Dinosaurs Using Laser Imaging and 3D Computer Modelling, PLoS ONE, (2009) 4 (2): e4532 doi:10.1371/journal.pone.0004532.

Gatesy, S. M., Baker, M. and Hutchinson, J. R   Constraint-Based Exclusion of Limb Poses for Reconstructing Theropod Dinosaur Locomotion. J. Vert. Paleo. (2009) 29: 535-544.

Hutchinson, J. R. and Garcia, M., *Tyrannosaurus* was not a fast runner. Nature, (2002) 415: 1018-1021.

Hutchinson, J. R. Biomechanical modeling and sensitivity analysis of bipedal running. I. Extant Taxa. J. Morph., (2004a) 262: 421-440

Hutchinson, J. R., Biomechanical modeling and sensitivity analyis of bipedal running ability. II. Extinct taxa, J. Morph., (2004b) 262: 441-461

Hutchinson, J. R., Anderson, F. C., Blemker, S. S., and Delp, S. L. Analysis of hindlimb muscle moment arms in Tyrannosaurus rex using a three-dimensional musculoskeletal computer model: implications for stance, gait, and speed. Paleobiology (2005) 32: 676-701.

Hutchinson, J. R. Ng-Thow-Hing, V., and Anderson, F. C. A., 3D interactive method for estimating body segmental parameters in animals: Application to the turning and running performance of *Tyrannosaurus rex.*   J. Theor. Bio. (2007) 246: 660-680.

Hutchinson, J.R., Bates, K., T., Molnar, J., Allen, V., and Makovicky, P. J., "Computational analysis of limb and body dimensions in *Tyrannosaurus rex* with implications for locomotion, Ontogeny, and Growth, PlosOne (2011) 6: e26037(1-20).

Sellers, W. I. and Manning, P. L., Lyson, T., Stevens, K. & Margetts, L. Virtual Palaeontology: Gait Reconstruction of Extinct Vertebrates Using Hight Performance Computing, *Palaeontologia Electronica,* (2009) 12.3.13A: 1-14.

Chapter 9   Casting a doubt for fast running ability of *T.rex*

Chapter 10   Dynamical calculation of the locomotion of *T.rex*

In the previous chapter, the author re-calculated required muscle mass for one leg based on the static theory. Theoretically, static theory tells nothing on running speed. It estimates required muscle mass for given weight. Speed factor lies outside the theory, namely, Froud number discussion, $Fr = \dfrac{v^2}{gh}$. This argument is said to be primitive on speed evaluation, which will be discussed in Chap.11. The work involving dynamics of the motion predicts how fast *T.rex* could run, theoretically. In this chapter, the author presents dynamical calculation of *T.rex* running motion using computer simulation.

10.1   Dynamical calculation of the locomotion of *T.rex*

Running motion is a periodic one, hence, expressing time change of each joint angle by Fourier expansion series is appropriate. Validity of this method was checked in advance on human locomotion. The motion capture of human running motion was accomplished by the combination of optical measurements and the use of force plate on the ground. These data were analyzed by reliable system VICON (Vicon Motion Systems). Next, time change of each joint angle is expressed by Fourier expansion series. Convergence within 1 % accuracy is checked by taking into account of 5th order Fourier expansion. Thus, an expression of 5th order Fourier expansion is a good method to describe periodic motion of each joint. For *i*-th joint angle $\theta_i(t)$ the expansion is expressed as follows,

$$\theta_i(t) = a_i^0 \sin(0t + \delta_i^0) + a_i^1 \sin(\omega t + \delta_i^1) + \cdots + a_i^5 \sin(5\omega t + \delta_i^5) \ , \tag{10.1}$$

where a_i^j, δ_i^j are the amplitude and the phase of *j*-th order of expansion series for *i*-th angle, respectively. The ω is angular velocity. The segment structure of *T.rex* is the same of Hutchinson and Garcia's model shown in Fig. 9.1 (a).

To study time-dependent dynamics, solid object model is used to describe the motion of *T.rex* limb. Namely, the model *T.rex* moves as one solid object for the external force $\vec{F}(\vec{r})$ as the following equations,

$$m_{body}\frac{d^2\vec{X}}{dt^2} = -m_{body}g\big|_y + \vec{F}(\vec{r}) \quad , \tag{10.2}$$

$$I\frac{d^2\Phi}{dt^2} = \vec{r}\cdot\vec{F}(\vec{r}) \quad , \tag{10.3}$$

where $\vec{X}$ and Φ are the position vector of the center of mass and the rotational angle of the object, respectively. The calculation is achieved in the sagittal plane, i.e., two dimensional space x (horizontal) and y (vertical). I, g and $\vec{r}$ are the momentum of inertia of T.rex, gravitational constant, and the position vector to the point of the force, respectively. The term $-m_{body}g\big|_y$ expresses that gravitational force acts in vertical direction y. The value of inertia I is chosen as I=19000 kg·m² in our work. Note that Hutchinson et al.'s value is I_{zz} =19200 kg·m² for 6583 kg T.rex, where I_{zz} is the inertia around the axis perpendicular to the sagittal plane (Hutchinson et al. 2007). Bates et al.'s value is 18890.29 kg·m² for "HAT" (Head-Arms-Torso) of 6071.82 kg T.rex (Bates et al. 2009). The both studies used the same specimen T.rex MOR555, however, difference of reconstruction leads to the slightly different estimation of the inertia. Our value is set close to these values. A solid object model is simple, however, it is known to express dynamics of moving object with many degree of freedom (Usami et al. 1998).

For calculating joint torque, or moment of the force about joint, a free-body diagram analysis is applied as shown in Fig.9.1 (b). For example, let call the foot segment as segment "1", and define the mass and the moment of inertia as m_1 and I_1, respectively. Then, the equations of motion for translation and rotation become as follows in (x, y) plane,

$$\vec{F}_1 - \vec{F}_2 - m_1 g\big|_y = m_1\vec{a}_1 \quad , \tag{10.4}$$

$$\vec{x}_{1g}\times\vec{F}_1 - \vec{x}_{2g}\times\vec{F}_2 + M_1 - M_2 = I_1\dot{\omega}_1 \quad , \tag{10.5}$$

where $\vec{F}_1$, $\vec{F}_2$ and $\vec{a}_1$ are the force from downside segment, the force from upper segment and acceleration, respectively. For rotational motion, $\vec{x}_{1g}$ and $\vec{x}_{2g}$ are the vector from the center of mass of 1-th segment to the point acting force $\vec{F}_1$ and $\vec{F}_2$. M_1 and M_2 are the moment of force between 0-th and 1-th, 1-th and 2nd segment, respectively. For the case of 1-st segment $\vec{F}_1$ corresponds to the ground reaction force. $\dot{\omega}_1$ is time derivative of angular velocity. Putting known

Chapter 10 Dynamical calculation of T.rex running motion

terms $\vec{F}_1$, $\vec{a}_1$, $\dot{\omega}_1$ and M_1 into Eq.10.4 yields unknown terms $\vec{F}_2$ and M_2. Thus, the moment of force acting to upper segment is obtained, subsequently.

A solid object approximation for the motion of the whole body. For the motion of whole body, solid object approximation is introduced. In the expression the total mass M, the moment of inertia I and gravitational constant g are, 6071 kg, 9.80 m/s², 19000 kg·m², respectively. The external force $\vec{F}(\vec{r})$ is ground reaction force (GRF), which acts along vertical direction y as the following relation,

$$F_y(y) = -ky - \gamma v_y \qquad , \qquad (10.6)$$

where y and v_y are the depth from the horizontal level and vertical velocity, respectively. This relation is composed by Hooke's law with spring constant $k=1.0 \times 10^7$ N/m and dumping term with coefficient $\gamma = 2.0 \times 10^5$ Ns/m in our simulation. This model gives appropriate solution of running motion with wide range of parameters k and γ in our simulation.

10.2 Evolutionary computation method

Searching the optimal Fourier coefficient δ_i^j for running motion is the next task. The other parameters are fixed in the simulation. Computational method for obtaining the optimal solution in many degree of freedom is usually not an easy task. So, a variety of approximation method has been proposed in many research areas. One of the famous and well studied method is genetic algorithm (GA) (Fraser 1970; Holland 1975; Goldberg 1989). A vast number of studies have been published in many research areas concerning to GA. This method is based on the idea of gene evolution observed in actual life systems. In this method digitized virtual genes are introduced, and its evolution is simulated. The virtual gene falls into a stable state in which the value of evaluation function has a local minimum. On the contrary, the introduction of virtual genes is not necessary for the present study. So, looking for another convenient approximation method is appropriate.

Another approximation method for obtaining near optimal solution is evolutionary computational method (Sellers and Manning 2007; Usami 1998; Fogel 1995). This method is known, as well as genetic algorithm method on searching near optimal solution. This method does not introduce digitized virtual gene, but change parameters of the system directly. Parameters converge into the local minimum rapidly, and the result is usually satisfactory. Then, we use

Chapter 10 Dynamical calculation of *T.rex* running motion

evolutionary computational method on this problem.

At first, we create several typical patterns of running motion at hand using 3D software 3dsmax (Autodesk co.). The typical patterns include various motions from flexed one to upright one. Next, we apply dynamical simulation described above. At the first stage of evolutionary computation, the model *T.rex* usually falls on the ground in the simulation space. Then, the parameters are improved by evolutionary computation method. The original set of parameters is slightly changed within a certain range, randomly. These sets of parameters are considered as children of the parent. Running motion having slightly changed parameters is calculated, and the best performance child is selected among children. Again, the parent of the best parameter set brings children who have slightly different values from the parent. Thus, near optimal solution on running motion is obtained as a result of evolutionary computation scheme.

In the simulation, there are many choices of the evaluation function to obtain appropriate solution. We have tried many types of function for obtaining running motion of the segment model of *T.rex*. As a result, the choice of the product of ground reaction force and forward velocity is suitable for the evaluation function on this problem. This is mainly due to the fact that legs of the segment model inevitably rotate around each joint, which generates driving force to move any of direction. Then, taking this condition yields smooth running motion of the segment model of *T.rex*.

10.3 Simulation Results — *Running Motion Generated by Evolutionary Algorithm*-

A typical running motion generated by evolutionary algorithm is shown as stick diagram in Fig. 10.1. Stride length of this motion is 6.2m. Cyclic period is 0.693 second, which leads to running speed of 8.9 m/s. This result is relatively similar to the observed data reported by Farlow (Farlow 1981), and Day et al. (Day et al., 2002).

Figure 10.2 displays stick diagram of the leg while the foot touches the ground. Hutchinson and Garcia's best guess mid-stance posture is shown as thick dashed line in Fig. 10.2(b) (Hutchinson and Garcia, 2002). It is noticed that Hutchinson and Garcia's posture is rather flexed than our upright posture. Note that zero moment point (ZMP) in bipedal locomotion usually moves from the heel side to the tip side of the foot during stance phase. The ZMP is usually defined in robotics as the one where total inertia force equals to zero. The ZMP is expressed as the triangle in Fig. 10.2 (b).

Chapter 10 Dynamical calculation of *T.rex* running motion

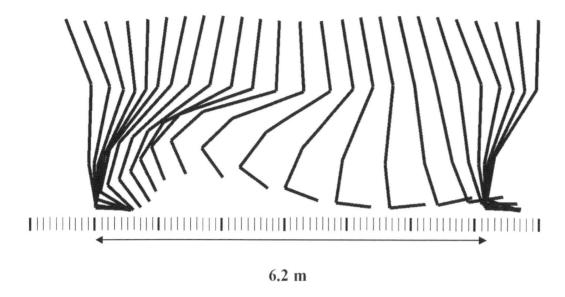

Fig.10.1 Stick diagram of the segment model of *Tyrannosaurus* limb in running motion. Running speed in this example is 8.9 m/s. Stride length is 6.2 m. and cyclic period is 0.693 s.

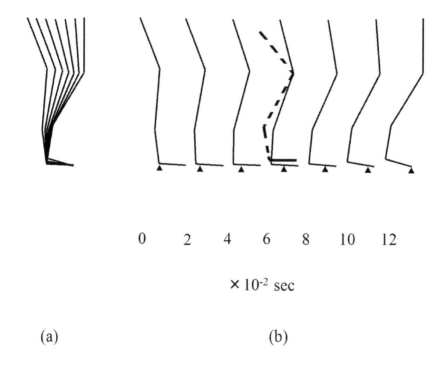

(a) (b)

Fig.10.2 Stick diagram of the segment of the limb while the foot touches on the ground. (a) is an ordinary stick diagram, and (b) is snapshot of 0.02 s interval. Thick dashed line in (b) represents Hutchinson and Garcia's mid-stance posture (Hutchinson and Garcia, 2002). Triangle shows zero moment point (ZMP) used for the calculation of required extensor muscle mass m_i in Eq.9.2.

Chapter 10 Dynamical calculation of *T.rex* running motion

Figure 10.3 shows vertical acceleration and extensor muscle mass of each joint at each moment during stance phase. For this calculation, the ground reaction force F_1 i.e., $-1 \times$ total mass $\times$ vertical acceleration is inserted into Eq.9.2, which yields the moment of force, or the torque of i-th joint M_i. Acting point of the ground reaction force F_1 in Eq.9.2 is the ZMP shown as the triangle in Fig. 10-2(b). Inserting M_i into Eq.9.2 yields the extensor muscle mass m_i which is expressed as a fraction to the total mass of *T.rex*.

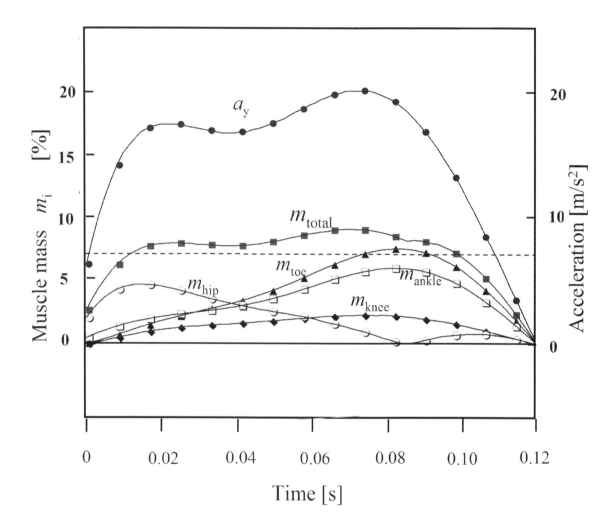

Fig.10.3 Required muscle mass m_i for i-th segment of the leg and vertical acceleration of the center of mass. The m_i is calculated using Eq.(9.1) and Eq.(9.2) while a foot touches on the ground. Black triangle, white square, black rhombus and white circle represent m_i of toe, ankle, knee and hip, respectively. Black square is the sum of hip, knee and ankle. Black circle represents vertical acceleration of the center of mass. The dashed line represents 7 %, which is a value that if required muscle mass for a joint exceeds this, *T.rex* could not run quickly in Hutchinson's framework (Hutchinson, 2004). Note that σ is set as 30 N/cm^2 in this graph.

Chapter 10 Dynamical calculation of *T.rex* running motion

In the calculation, the center of mass is located at hip joint. The problem of the center of mass is discussed in detail at the later section. In Fig. 10.3, the vertical acceleration shows two local maximum. The first local maximum corresponds to the event that the heel of the foot touches the ground. On the other hand, the second local maximum corresponds to the event that the tip of the foot kicks on the ground. By this event body jumps up, and goes forward. The maximum value of vertical acceleration is observed as 20 m/s². In Hutchinson et al's work, this value is set as 2.5× gravity~24.5 m/s² for 20 m/s running (Hutchinson and Garcia, 2002; Hutchinson, 2004). Hence, our value of the maximum acceleration is slightly smaller than the one of their work. This effect contributes to decrease required muscle mass, which leads to the increase of probability of *Tyrannosaurus* running. Recent work of Gatesy et al. presented that 1.87× gravity vertical acceleration would be achieved within allowed parameters range (Gatesy et al., 2009). Our value is rather close to the one of 2009's work. Note that the parameters used in the static evaluation method for postures is re-examined in this work. Then, detailed discussion is given in the discussion section.

In Fig. 10.3, a fraction of muscle mass for *i*-th joint m_i is calculated by Eq.(9.2), and the sum of m_i, i.e., $m_{\text{total}} = \sum m_i$ are also shown. As is Hutchinson's work m_i of toe joint is omitted (Hutchinson, 2004). because the ankle extensors could have been producing most of the required toe joint moments. Note that the value of σ is set as 30 N/cm² in this evaluation, which is also shown in Table 10.1 together with the other parameters. In Hutchinson's criterion if m_i surpasses 7%, the bipedal animal is less likely to run quickly. It is observed in Fig. 10.3 that each m_i is within this criterion except for m_{toe}. The m_{toe} exceeds 7 %, however, Hutchinson's criterion omit this in running evaluation. Then, at this point it is difficult to conclude that *Tyrannosaurus* of 6071 kg could not run at least in a speed of 8.9 m/s

Comparing time change of each quantity with well studied extant animal data, several features have been drawn attention. At first, the maximum value of vertical acceleration 20 m/s² is quite similar of extant bipedal animal. Second, the overall behavior of the time change of each m_i is also relatively similar to the extant bipedal, for example, human's one. However, appearance of two local maximum of vertical acceleration is not the same feature with extant bipedal data. In the well studied human locomotion, vertical acceleration is almost changed as sine curve with time. This is mainly due to the fact that each part of body is smoothly connected. And, it prevents sudden change of acceleration. Our *T.rex* model is only made by legs part, then there is no room to absorbs large change of acceleration. Each body part such as neck, trunk and tail would absorb such large change of acceleration in *T.rex*. Then, we correct this discrepancy by giving sine function for vertical acceleration in the calculation of Fig. 10.3.

The result is shown in Fig. 10.4, in which the maximum value of acceleration is set equal to the one of Fig. 10.3. Namely, we give sine function for vertical acceleration, and re-calculate m_i. In Fig. 10.4 time change of each m_i is almost similar to the one of Fig. 10.3. For the case of Fig. 10.4 the sum of m_i has one maximum corresponding the fact that the ground reaction force has one maximum. We notice that the maximum value of $m_{total} = \sum m_i$ is almost the same of Fig. 10.4, i.e., m_{total} =9.5 %. And the position of the peak of total m_{total} moves to the center of time duration than Fig. 10.4, which does not affect to the maximum value of each m_i. In this case each m_i does not exceed 7 %. It means that T.rex could run in Huchinson's criterion (Hutchinson, 2004), which is appeared in our running simulation.

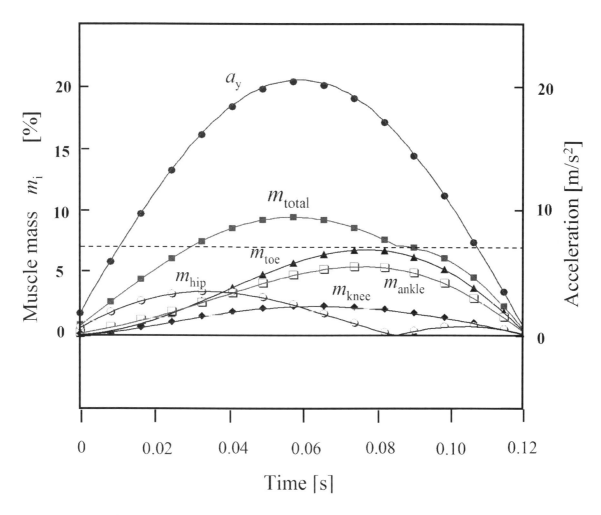

Fig. 10.4   The same of Fig. 10.3 in the case of vertical acceleration being sine curve. Vertical acceleration is corrected as sine function with time as black circle. Required muscle mass m_i is re-calculated, and shown.   Each required muscle mass m_i does not exceed the line of 7 %.

Chapter 10 Dynamical calculation of *T.rex* running motion

10.4 Obtained parameters in running simulation

In this subsection values of obtained parameters in our running simulation are summarized. Fourier coefficient of each joint for running motion is expressed as Eq.10.1. The optimized parameters are listed in Table 10.1. The amplitude a_i^j is expressed as the degree, and the phase δ_i^j is expressed as radian. The zero-th order value of a^j (j=0) represents central position of swing angle of each leg. Basically, simulation is accomplished as the condition with the fixed value of a_i^j. However, a_2^0 i.e., the central position of swing angle of the knee is included in optimization parameter with the range of ±5 degree. This introduces the increase of optimization space to bring appropriate running motion. Thus, optimized running motion is searched with the change of phase δ_i^j and a_2^0. In Eq.10.1 and Table 10.2 θ_4 and θ_2 are defined as an angle from vertically downward direction to forward. θ_3 is defined as an angle from vertically downward direction to backward. θ_1 is defined as an angle from horizontal line to upward. Evolution algorithm is accomplished as follows; Choosing four parameters among δ_i^j randomly. Next, random value between $(-\pi \sim \pi)/5j$ is added for the original δ_i^j, where j is an order. The division by j is accomplished because large phase variance of higher order term induces inappropriate motion. The motion having the lowest of the product of ground reaction force and forward velocity is selected among 10000 trials of this random algorithm. This calculation is repeated until the evaluation function is converged.

Year	Autor	Parameter	Hip	Knee	Ankle	Toe	
2009	Gatesy *et al.*		0.38	0.14	0.09	0.070	
2005	Hutchinson		0.38	0.14	0.09	0.10	
		r	0.42-0.21	0.2-0.14	0.11-0.07		
			0°~40°	0°~80°	0°~−80°		
2004	Hutchinson		0.37	0.22	0.12	0.070	
2002	Hutchinson and Garcia		0.37	0.22	0.11	0.065	
2009	Gatesy *et al.*		0.85	0.40	0.26	0.18	
2004	Hutchinson	L	0.85	0.4	0.26	0.18	
2002	Hutchinson and Garcia		1.2	0.52	0.39	0.19	
2004	Hutchinson	θ	0	0	0	0	
2002	Hutchinson and Garcia		9	21	22	21	
	This work	r	0.38	0.2-0.14	0.11-0.07	0.070	Fig.10.3-4 12.2-5, 13.1-2
				0°~80°	0°~−80°		
		r	0.37	0.22	0.12	0.070	Fig.12.1
		L	0.85	0.40	0.26	0.18	Fig.10.3-4 12.1-5, 13.1-2
		θ	0	0.0	0	0	All Simulations
		σ	σ =30 N/cm²				All Simulations

Chapter 10 Dynamical calculation of *T.rex* running motion

Table 10.1. Summary of parameters in the literature and this work.

Order	Hip a_1	Knee a_2	Ankle a_3	Toe a_4	Hip δ_1	Knee δ_2	Ankle δ_3	Toe δ_4
0	4.542	37.363	20	−12.061	1.571	1.571	1.571	1.571
1	15.138^4	23.215^3	10.013^2	11.565^1	−3.4964^4	−0.9907^3	−0.3615^2	3.0641^1
2	5.028	17.088	0	12.159	1.9479	1.6191	0.6573	0.7816
3	1.871	2.409	0.003	2.758	2.6282	3.0033	0.5829	1.2058
4	0.182	1.005	0.003	1.183	1.0328	3.2611	−0.2863	0.9168
5	0.616	0.739	−0.004	0.88	−2.5246	−2.2349	0.1085	−2.8568

Table 10.2. The optimized parameters in Eq.(10.1)..

Summary

1 Running motion in a speed of 8.9 m/s is obtained by numerical simulation. The m_{total} which expresses a fraction of muscle mass to the total body mass obtained by the calculation is 9%.

2 The m_{total} of mass distribution analysis revealed that the m_{total} is 16.0 % for MOR555 (Bates et al. 2009), 14.2 % for MOR 555 (Hutchinson et al., 2007), 14.4 % for BHI3033 (Bates et al. 2009).

　New detailed computer-aided mass analysis shows that a range of the ratio of extensor muscle masses to the total body mass is 14.6%~25.4% for four adult specimens (Hutchinson et al., 2011). Table 10.3 summarizes the detail.

3 The center of mass is located in the calculation. If it is 36cm cranial that is a value obtained in the work (Hutchinson et al. 2007), the value of m_{total} is corrected as twice larger one, which will be discussed in Chap. 12 in detail. However, predicting the location of the center of mass is a difficult task, because the value of the m_{total} is very sensitive to it, which will be

Chapter 10 Dynamical calculation of *T.rex* running motion

discussed in Appendix.

4   The value of maximum muscle stress σ is set as 30 N/cm². The value of m_{total} depends on σ, greatly. The issue relating σ was discussed in Chap.4.

	CM 9380	FMNH PR 2081	BHI 3033	MOR 555	BMR P2002.4.1
Min	14.6	15.4	20.5	20.1	20.6
Max	15.5	17.6	25.4	23.7	28.8

Table 10.3 Ratio of extensor mass of a limb to the body mass (%)   (Hutchinson, et al., 2011).
A specimen BMR P2002.4.1 is supposed to be a juvenile with mass of 639~1269 kg.

References

Bates, K. T., Manning, P. L., Hodgetts, D., and Sellers, W. I. Estimating Mass Properties of Dinosaurs Using Laser Imaging and 3D Computer Modelling, PLoS ONE 4 (2), (2009) e4532 doi:10.1371/journal.pone.0004532.

Day, J. J., Norman, D. B., Upchurch, P., and Powell, H. P. Dinosaur locomotion from a new trackway, Nature, (2002) 415: 494-495.

Farlow, J. O. Estimates of dinosaur speeds from a new trackway site in Texas. Nature, (1981) 294: 747-748.

Fogel, D. B. Evolutionary Computation, Toward a New Philosophy of Machine Intelligence (IEEE Press, Piscataway, NJ, 1995).

Hutchinson, J. R., and Garcia, M. Tyrannosaurus was not a fast runner. Nature, (2002) 415: 1018-1021.

Hutchinson, J. R., Biomechanical modeling and sensitivity analyis of bipedal running ability. II. Extinct taxa, J. Morph., (2004) 262: 441-461

Hutchinson, J.R., Bates, K., T., Molnar, J., Allen, V., and Makovicky, P. J., Computational analysis of limb and body dimensions in Tyrannosaurus rex with implications for locomotion, Ontogeny, and Growth, PlosOne, (2011) 6: e26037(1-20).

Gatesy, S. M., Baker, M. and Hutchinson, J. R., Constraint-Based Exclusion of Limb Poses for Reconstructing Theropod Dinosaur Locomotion. J. Vert. Paleo., (2009) 29: 535-544.

Hutchinson, J. R. Ng-Thow-Hing, V., and Anderson, F. C. A 3D interactive method for estimating body segmental parameters in animals: Application to the turning and running performance of Tyrannosaurus rex, J. Theor. Bio., (2007) 246, 660-680.

Sellers, W. I. and Manning, P. L., Lyson, T., Stevens, K., and Margetts, L. Virtual Palaeontology: Gait Reconstruction of Extinct Vertebrates Using Hight Performance Computing, Palaeontologia Electronica, (2009) 12.3.13A: 1-14.

Usami, Y., Hirano, S., Inaba, S., and Kitaoka, M. Reconstruction of Extinct Animals in the Computer. Reconstruction of Extinct Animals in the Computer. in (Adami, C., Belew, R. K., Kitano, H. & Taylor C.E. eds.) pp.173-177 (Artificial Life VI, MIT Press, UCLA, 1998).

Chapter 11 Maximum running speed of *T.rex*

11.1 A relation of our work with preceding study

In this chapter, we discuss on running speed of *T.rex* based on our simulation result quantitatively. Sellers et al. presented 8.0 m/s running speed in Table 1 of the reference (Sellers et al., 2009). Its cyclic time was 1.199 s, and the stride length was 9.559 m. We have tried to search running motion with cyclic time of 1.2 s, however, we have failed to obtain appropriate solution. Such slow period brings large ground force. The vertical acceleration reaches 97 m/s², which leads G factor as 9.9, namely 9.9×9.8=97 m/s². We show a sample of our simulation result in Fig. 11.1.

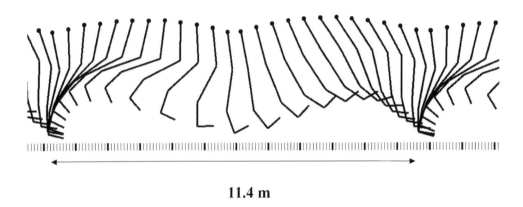

11.4 m

Fig. 11.1 A sample of our simulation result based on the parameters close to Sellers et al's [5]. Cyclic time is 1.2 s, which leads to large up-down motion. This vertical acceleration reaches 97.2 m/s². The stride length is 11.4 m, and the running speed is 9.5 m/s.

In this case, stride length is 11.4 m, and cyclic time is 1.2 s, which yields 9.5 m/s running. Large up-down motion of the hip joint is observed in Fig. 11.1, which generates large vertical acceleration. Sellers et al. also showed the results of considerably high speed up to 57 m/s, however, the other data such as vertical acceleration was not described. Then, we set cyclic time as 0.69 s at the starting point of numerical simulation. And we increase the amplitude of leg, and slightly increase cyclic time to investigate the change of running speed and stride length.

Chapter 11   Maximum running speed of *T.rex*

11.2   What is the maximum running speed of *T.rex*?

The most important question throughout this book is a question what is the maximum running speed of *T.rex*?   In this chapter, the author's simulation result is presented. However, it should be noted that the result depends on simulation condition. In this and next chapter the author studies on uncertainty of parameters involved in the theoretical framework in detail. For example, our model of *T.rex* has segments with length of 1.13, 1.26, 0.699 and 0.584 m for thigh, shank, metatarsus and foot, respectively, that are identical with Hutchinson et al.'s works (Hutchinson and Garcia, 2002; Hutchinson, 2004b; Gatesy et al., 2009). If these values slightly change, the result may be different from the one that we describe in this section. Ideally, theoretical researcher should estimate all parameter dependence in the work. However, the present study does not fully investigate all parameter dependence. Then, the result in this section is said to be an example of possible simulation studies.

We have calculated huge number of running simulation trials. Generally, vertical acceleration increases with running speed. However, its tendence is not uniform according to many different running patterns. The increase of minimum vertical acceleration is observed with the increase of running speed. Two or three times larger than gravitational constant 9.8 m/s² is plausible for animal running motion. The detailed result is appeared in our separate paper (Usami, 2014). The maximum running speed appeared in our running simulation is 14.1 m/s with vertical acceleration 28.3 m/s² whose stride length is 10.1 m, and cyclic period is 0.716 s. For higher running motion vertical acceleration shows considerably large value, for example, 20 m/s running brings 60 m/s² vertical acceleration.

Required muscle mass m_i of 14.1 m/s running during stance phase is calculated. The m_i expresses a fraction of muscle mass for *i*-th joint to the total body mass. Calculation method for m_i is described in Chap.10. Hutchinson stated that if m_i surpasses 7%, the bipedal animal is less likely to run quickly (Hutchinson, 2004). The detailed result is appeared in our separate paper (usami, 2014). Here, we state that none of our value of m_i does not exceed 7%. Then, the result of our numerical simulation suggests a possibility of quick running of *T.rex*. The sum of m_i, i.e., $m_{total} = \sum m_i$ is 9.2 at the maximum. In this evaluation m_i of toe joint is omitted (Hutchinson, 2004), because the ankle extensors could have been producing most of the required toe joint moments. Recent computer aided mass distribution analysis revealed that the m_{total} is 14.2%~16.0% (Bates et al., 2009; Hutchinson et al., 2007). Then, the result of our running simulation supports a possibility of quick running of *T.Rex* even in this criterion.

Note that the center of mass is located at the hip joint in the calculation. If it is 36cm cranial

Chapter 11 Maximum running speed of T.rex

that is a value obtained in the work (Hutchinson et al. 2007), the value of m_{total} is corrected as approximately twice larger one. Detailed discussion on the problem of the center of mass is given in Chap.12 and Chap.15.

The maximum value of vertical acceleration appeared in simulation is 28.3 m/s². Concerning to Gatesy et al's work published in 2009, 18.3 (~1.87×9.8) m/s² vertical acceleration is allowed in their estimation on static postures. However, they also stated as follows: "In light of the large number of options available to most theropods at running GRFs of 2-4 BW (Body Weight), further optimization analysis and consideration of the entire stride cycle may reveal why specific poses are chosen over so many alternatives" (Gatesy et al., 2009). Our result of 29.3 m/s² (~3.0 BW) is within this range. And, we obtained entire stride cycle by dynamical calculation with well described parameters. Then, our work is a possible answer to their unsolved question.

summary

1. The m_{total} which expresses a fraction of muscle mass to the total body mass obtained by the calculation is 9.2%. (The m_{total} of mass distribution analysis is 14.2%~16.0% (Bates et al., 2009; Hutchinson et al., 2007)).

2. The center of mass is located at the hip joint in the calculation. If it is 36cm cranial that is a value obtained in the work (Hutchinson et al. 2007), the value of m_{total} is corrected as twice larger one. (Related discussion is given in Chap.12 and Chap.15.)

3. The value of maximum muscle stress σ is set as 30 N/cm². The value of m_{total} also depends on σ, greatly. The issue relating σ was discussed in Chap.4

11.3 A relation between relative stride length and relative velocity

A relation between relative stride length $\hat{L} = L_{st}/h$ v.s. relative velocity $\hat{V} = v/\sqrt{gh}$ is plotted in Fig. 11.2, where L_{st}, h, v, g are stride length, hip height, velocity, and gravitational constant, respectively. Our data is plotted as triangle with the dot. The other data are various birds and human taken from the reference (Gatesy and Biewener, 1991).

Chapter 11 Maximum running speed of *T.rex*

One of characteristics of ours is low relative stride length and high relative velocity. One reason that our data gives low relative stride length lies in the assumption of the value of the hip height. We assume hip height as $h = 3.1$ m, which is similar to Sellers et al's assumption (Sellers et al., 2007). However, Hutchinson assumed this value as $h = 2.5$ m. If we assume $h = 2.5$ m, our data becomes slightly closer to the other observed data of birds etc.

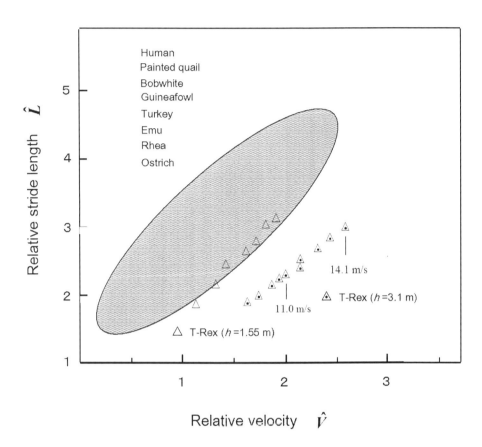

Fig. 11.2 A graph of relative stride length against dimensionless velocity. White triangles with dot are our simulation data of *T.rex* running in which hip height h is 3.1 m. White triangle is data of *T.rex* running with hip height of 1.55 m. The other data are of human and various birds which are shown as dark area, those are shown in the reference (Gatesy and Biewener, 1991).

Chapter 11   Maximum running speed of *T.rex*

Another reason to show low $\hat{L}$ is considered to be large limb structure of *T.rex*. For verifying this idea, we accomplish simulation of small *T.rex* running. We create small *T.rex* with hip height h =1.55 m, and accomplish simulation with the same condition. The result is shown as the triangle with letter of *T.rex* (h=1.55 m). The data becomes to be close to the other data for this case. Thus, the reason giving low $\hat{L}$ for *T.rex* running motion appeared in our simulation is considered to be the large structure of the limb. It is noted that the linear relation between relative stride length and relative velocity is in agreement with the other data.

11.4   On the Froude number discussion

As is noted at Chap.8, an estimation of running speed using Froude number $Fr = v/^2 gh$ involves large uncertainty. In Fig. 11.2 relative stride length $\hat{L} = L_{st}/h$ has roughly linear relation with relative velocity $\hat{V} = v/\sqrt{gh}$, however, the data is scattered. For a certain value of relative stride length $\hat{L}$, nearly two fold variance of relative velocity $\hat{V}$ is observed. Although ostrich's data denoted by black circle shows convergence rather than two fold variance, it is not certain in what extent *T.rex* locomotion resembles to ostrich's one. The discussion of Hutchinson et al. is as follows (Hutchinson and Garcia, 2002). Observed ostrich's running data at 12 m/s gives a value of *Fr*=16 with a multiply factor to gravity G=2.7. They calculated that *T.rex* could not hold mid-stance posture of running with the force of G=2.5. Hence, *T.rex* could not run with speed of *Fr*=16. In the calculation, employed value of hip height h is 2.5m, which leads to running velocity as 20 m/s, i.e., $16 = 20^2 / 9.8 \cdot 2.5$. Then, they concluded that *T.rex* could run with speed of 20 m/s. However, it is known to be difficult to predict running speed using Froude number, accurately. For example, if we assume that the total length of *T.rex* leg is 3.1 m, Froude number becomes 13 instead of *Fr*=16, i.e., $13 = 20^2 / 9.8 \cdot 3.1$. Actually, Sellers et al. and I used similar value of hip height as h=3.089 m and h=3.1 m, respectively.

Let us discuss on Froude number for large bipedal dinosaur running in detail. In our simulation Froude number is calculated as Fr=2.7~6.5 for v=8.9 m/s ~ 14.1 m/s running. These are shown as the white triangle in Fig. 11.3. Black triangle shows Sellers et al.'s data of Fig. 4 in the reference (Sellers et al., 2007). Black square shows the data of Hutchinson et al's for the case of h=2.5 m (Hutchinson and Garcia, 2002).

Chapter 11 Maximum running speed of *T.rex*

From fossilized foot-print, Farlow reported L_{st}=6.59 m dinosaur trackway from the Lower Cretaceous of Texas U.S. (Farlow, 1981). He reported that stride/hip height ratio was 4.3. From the data, hip height and Froude number are calculated as h=1.50 m, and Fr=8.39, respectively. Day et al. reported dinosaur trackway of L_{st}=5.65 m in 163 million years old strata in U.K (Day, et al., 2002). Hip height is estimated as h=1.93 m from footprint, and reported that running speed might be v=8.11m/s. Corresponding Froude number is calculated as Fr=3.5 in this case. These observation data are shown as black and white circle in Fig. 11.3.

It seems from the data that Hutchinson et al's assumption of Fr=16 would be large value for *T.rex* running. It is noted that Sellers et al. presented the data that required muscle mass is 22.5 % for Fr=3.8 running in a speed of 10.7 m/s (Sellers and Manning, 2007). As is noted at Chap.8, Bakker and Paul proposed 20 m/s and 17.9 m/s running based on the morphological consideration of muscle and limb structure (Bakker, 1986; Paul, 1988). These are plotted as black and white rhombus in Fig. 15, respectively, with the assumption of h=3.1 m.

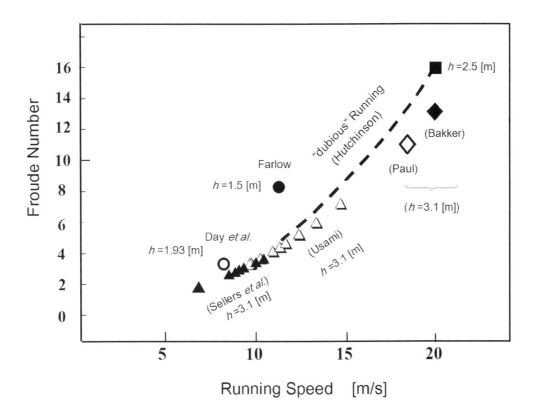

Fig. 11.3 Froude number v.s. running speed of dinosaur in the literatures and this study. Black and white triangles show simulation results by Sellers and Manning (Sellers and Manning, 2007) and us,

Chapter 11   Maximum running speed of *T.rex*

respectively. Black circle represents Farlow's data from fossilized foot-print with hip height $h=1.5$ m (Farlow, 1981).   White circle represents Day et al.'s data from foot-print with hip height $h=1.93$ m (Day et al., 2002).   Black and white rhombus are Bakker and Paul's estimation for 20 m/s and 17.9 m/s running, respectively (Bakker, 1986; Paul, 1988). All of these states that *T.rex* would run in this speed range. On the contrary, black square and thick dashed curve show the range of no and "dubious" running proposed by Hutchinson and Garcia (Hutchinson and Garcia, 2002) and Hutchinson (Hutchinson, 2004), respectively. Black square is the data of $Fr=16$ stated as running *T.rex* in a speed of 20 m/s with hip height $h=2.5$ m (Hutchinson and Garcia, 2002).

All of these works state that *T.rex* would run in the speed range of 7 m/s ~ 20 m/s, although the methodologies of them are different. On the contrary, only the works using the static method stated that *T.rex* would not run in a speed of 20 m/s (Hutchinson and Garcia, 2002). Hutchinson stated in the work of 2004 as " speeds >11m/s remains dubious" (Hutchinson, 2004). These speed regions are shown in Fig.11.6 as thick broken line. However, it is noted that speed estimation using Froude number is qualitative, and has uncertainty in quantitative evaluation.

On the reconstruction of locomotion of bipedal dinosaur from footprint, several researchers have thrown a question to use Alexander relation (Alexander, 1976) for speed estimation. To use this relation, hip height h must be estimated from foot length FL. Alexander's first proposal was $h=4FL$, namely, hip height is four times foot length (Alexander, 1976).   However, Thulborn stated that the factor ranges from 4.5 to 5.9 according to type and size of dinosaur (Thulborn, 1990). Recently, Rainforth et al. re-analyzed this factor using 24 specimens from different dinosaurian groups, and concluded that speed estimation could be incorrect by a factor of two (Rainforth and Manzella, 2007). They stated "there is no reliable way to estimate hip height from footprint length, either using morphometric or allometric equations " (Rainforth and Manzella, 2007). Thus, it should be kept in mind that discussion using Alexander relation and Froude number includes large uncertainty.

Chapter 11 Maximum running speed of *T.rex*

References

Alexander,R. Mc. N. 1976. Estimates of speeds of dinosaurs. Nature   261: 129-130.

Bakker, R. T. *Dinosaur Heresies* (William Morrow, New York, 1986).

Day, J. J., Norman, D. B., Upchurch, P. & Powell, H. P.   Dinosaur locomotion from a new trackway, Nature, (2002) 415: 494-495.

Farlow, J. O.   Estimates of dinosaur speeds from a new trackway site in Texas. Nature, (1981) 294: 747-748.

Gatesy, S. M. & Biewener, A. A.   Bipedal locomotion: effects of speed, size and limb posture in birds and humans, *J. Zool., Lond.,* (1991) 224: 127-147.

Gatesy, S. M., Baker, M. and Hutchinson, J. R.   Constraint-Based Exclusion of Limb Poses for Reconstructing Theropod Dinosaur Locomotion. J. Vert. Paleo., (2009) 29: 535-544.

Hutchinson, J. R. & Garcia, M. Tyrannosaurus was not a fast runner. Nature, (2002) 415: 1018-1021.

Hutchinson, J. R. 2004. Biomechanical modeling and sensitivity analyis of bipedal running ability. II. Extinct taxa, J. Morph., (2004) 262: 441-461.

Paul, G. S.   *Predatory Dinosaurs of the World* (Simon & Schuster, New York, 1988).

Rainforth, E. C. & Manzella, M. Estimating speeds of dinosaurs from trackways: a re-evaluation of assumptions. in (Rainforth E. C. ed.) Contributions to the paleontology of New Jersey (II), pp41-48 (GANJ 24, 2007).

Sellers, W. I. & Manning, P. L. Estimating dinosaur maximum running speeds using evolutionary robotics, *Proc. Roy. Soc.,* (2007) *B* 274: 2711-2716.

Thulborn, R. A.   *Dinosaur Tracks* (Chapman & Hall, London, 1990).

Usami, Y. Evolutionary computation strategy is superior than simulated annealing for obtaining running motion of dinosaur. Proceedings of 6th Intern. Conf. Evolutionary Comp. Theory and Application, Rome, (2014).

Chapter 12   A problem of the position of center of mass

Chapter 12   A problem of the position of center of mass

A problem of the center of mass is a critical problem on considering locomotion of *T.rex*. A value of required muscle mass m_i depends greatly on the position of the center of mass. In this section, we investigate this problem systematically.

12.1 CM of various posture

In Fig. 12.1 (a) we show contour map of $m_{\text{total}} = \sum m_i$ as the change of limb configuration and the position of the center of mass. The horizontal axis corresponds to the change of limb configuration as shown in Fig. 12.1 (b). The limb configuration denoted A is a posture of Hutchinson's best guess of running motion (Hutchinson and Garcia, 2002; Hutchinson, 2004; Gatesy, et al., 2009).

From A, limb configuration is smoothly changed through B, C, D and to E. B is a posture of mid-stance of our running motion shown in Fig. 10.2. D is a configuration in which each leg is aligned vertically except for foot segment.   E is the configuration of Hutchinson's lowest model giving the minimum $m_{\text{total}} = \sum m_i$ (Hutchinson and Garcia, 2002).

The vertical axis of Fig. 12.1 (a) shows the position of the center of mass. The base line is the case that the center of mass is located upon the heel of the foot. The upper line is the case that the center of mass is located upon the tip of the foot. Required total muscle mass $m_{\text{total}} = \sum m_i$ is plotted as contour with graded grey color. Gravitational constant and its multiplier are set as 2.5×9.8 for comparing Hutchinson and Garcia's work. At first glance, we notice the large change of m_{total}.

For the posture of A, m_{total} varies from 15 % (heel side) to 28 % (tip side). The center of mass of Hutchinson's model is denoted as triangle (Hutchinson, 2004), which yields m_{total}=21 %. In the case of D, the minimum value of m_{total} is 0 %, however, the maximum m_{total} reaches to 28 %. This increase of required muscle mass is said to be large. Interestingly, Hutchinson et al's posture (Hutchinson and Garcia, 2002) giving the lowest m_{total} denoted E shows the largest change of m_{total}. The value varies from 6 % to 36 %.

On the contrary, our running mid-stance posture denoted B gives the least variance of m_{total}. It varies moderately from 10 % to 20 %. Especially, it is interesting to observe broad flat region around m_{total}=10 %. Hence, it is said that this configuration has a special meaning concerning to the posture of *T.rex*. On the contrary, it may be said that the posture denoted E is rather unnatural, because, this gives the largest change of m_{total}.   Note that in Bates et al.'s estimation(Bates, et al.,

Chapter 12   A problem of the position of center of mass

2009), the position of the center of mass is in a range of 0.295-0.652 m cranial of the hip joint. Gatesy et al.'s hip position is located on the middle third of the foot length. And evaluating point of GRF is located between 0.4 m and 0.7 m cranial of hip joint (Gatesy et al., 2009). Hutchinson et al. examined wide variance of mass distribution about four adult and one juvenile *T.rex* in 2011 (Hutchinson et al., 2011). Among the data, the center of mass is obtained in a range of 0.376~0.572 m cranial of hip joint for the minimal and maximal mass model of the five specimens. These values are within our evaluation range.

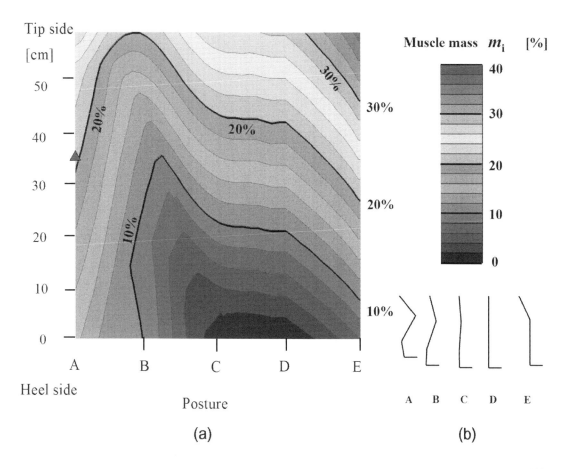

Fig.12.1 (a) represents two dimensional contour of required muscle mass $m_{total} = \sum m_i$ in different limb configuration (b). The limb configuration is changed from Hutchinson's mid-stance posture A through our mid-stance posture B, linear configuration D, and to Hutchinson's posture giving the lowest m_{total} E. Vertical axis represents the position of the center of mass. The baseline corresponds to the case that the center of mass is located upon the heel of the foot. The top line corresponds to the case that the center of mass is located upon the tip of the foot. Required muscle mass m_{total} is expressed as contour line with graded grey color. The triangle expresses the point of Hutchinson's best guess model (Hutchinson and Garcia, 2002; Hutchinson, 2004).

Chapter 12   A problem of the position of center of mass

12.2 CM of running motion

(i) m_{total}

To study the influence of the position of the center of mass, two dimensional contour map of m_{total} is also calculated for the case of our running motion as Fig. 12.2(a). Fig. 12.2(b) displays leg configuration in stance phase of Fig. 10.2. The horizontal axis of Fig. 12.2(a) corresponds to the change of leg configuration from A to E.

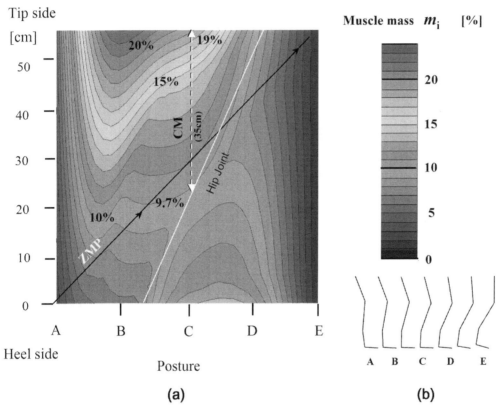

Fig.12.2 Two dimensional contour of $m_{total} = \sum m_i$ during stance phase in running motion.

Required muscle mass m_i during stance phase of Fig. 10.2 is expressed as 2-D contour map with graded grey color in (a). The limb configuration is shown in (b). Each configuration lettered as A, B, C, etc. corresponds to the horizontal axis of (a). Vertical axis represents the position of the center of mass. The baseline corresponds to the case that the center of mass is located upon the heel of the foot. The top line corresponds to the case that the center of mass is located upon the tip of the foot. Zero moment point (ZMP) is shown as the diagonal line from left-bottom to right-top. The position of the hip joint is also shown from mid-bottom to right-top. The center of mass position is located 0.36m cranial of the hip joint in the reference (Hutchinson and Garcia, 2002). Then, the width of 0.35 m is shown as

Chapter 12   A problem of the position of center of mass

dashed vertical arrow in Fig. 12.2(a), which is the closest value to 0.36m. In the observations of bipedal animal running, the point where the ground reaction force acts moves from the heel side to the tip side linearly in time. The change of ZMP (zero moment point) is shown as the diagonal line from left-downward to right-upward in Fig. 12.2(a). This movement of ZMP is shown as the triangle in Fig. 10.2(b). This map is calculated in the static condition, i.e., each $\vec{a}_i$ and $\vec{\omega}_i$ in Eq.(10.4), (10.5) is zero. Hence, only left-upward side of ZMP line is meaningful. The center of mass position is located 0.35m cranial of the hip joint in the reference (Hutchinson and Garcia, 2002). Then, the width of 0.35 m is shown as dashed vertical arrow, which is the closest value to 0.35 m appeared in Fig. 12.2(a).

(ii) m_{hip}, m_{knee} and m_{ankle}

Figures 12.3~12.5 are the same of Fig. 12.2, but for the value of hip, knee, and ankle joint, respectively. In Fig. 12.3 m_{hip} is in the range of 2.1 %~7.5 %, which is almost within the criterion of 7%. For knee joint shown in Fig. 12.4, m_{knee} is in the range of 3.5 %~2.0 %. Then, there is also no problem for quick running. However, for the case of ankle joint shown in Fig. 12.5 the range is 3.9 %~10.0 %.

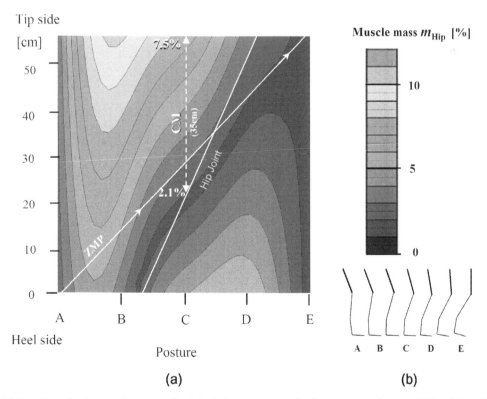

Fig.12.3   Required muscle mass for hip joint $m_i = m_{\text{hip}}$ during stance phase of Fig. 10.2. The m_{hip} in stance phase shows small value as $m_{\text{hip}} = 2.1$ % at the hip joint in C, and 7.5 % at 35 cm cranial.

Chapter 12　A problem of the position of center of mass

For this case m_{ankle} excesses the criterion of 7%. However, the excess amount is not so large. So we could not conclude immediately that *T.rex* could not run quickly, only from this result. Rather, almost region of Figs. 12.3~5 is covered in the range of 7 %. Then, the final assessment of running ability should be waited until the evaluation of the other factors is completed.

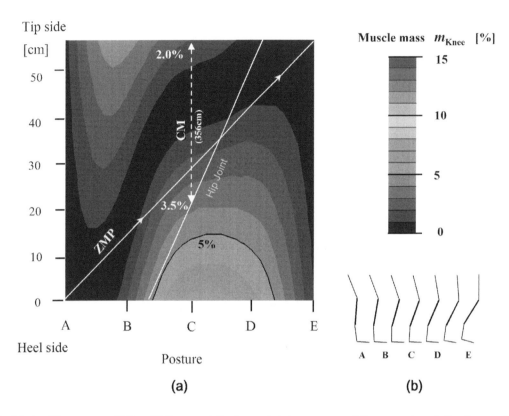

Fig.12.4　The same of Fig. 12.3 except for knee joint $m_i = m_{knee}$. The m_{knee} is relatively small as m_{hip} =3.5 % at the hip joint in C, and 2.0 % at 35 cm cranial.

Chapter 12   A problem of the position of center of mass

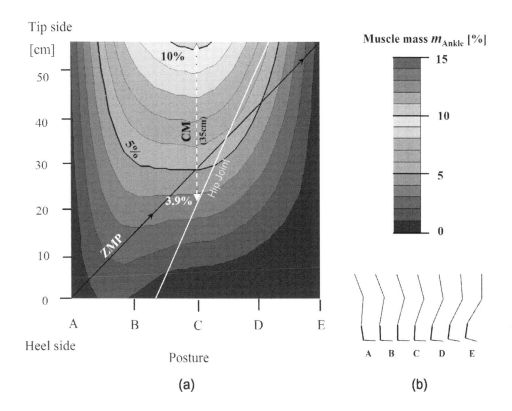

Fig.12.5 The same of Fig.12.3 except for ankle joint $m_i = m_{ankle}$. The m_{knee} is small at the hip joint in C as 3.9 %, however, it increases to 10 % at 35 cm cranial.

References

Bates, K. T., Manning, P. L., Hodgetts, D. & Sellers, W. I.   Estimating Mass Properties of Dinosaurs Using Laser Imaging and 3D Computer Modelling, PLoS ONE, (2009) 4 (2): e4532 doi:10.1371/journal.pone.0004532.

Gatesy, S. M., Baker, M. & Hutchinson, J. R.   Constraint-Based Exclusion of Limb Poses for Reconstructing Theropod Dinosaur Locomotion. J. Vert. Paleo., (2009) 29: 535-544.

Hutchinson, J. R. & Garcia, M. *Tyrannosaurus* was not a fast runner, Nature, (2002) 415: 1018-1021.

Hutchinson, J. R. 2004. Biomechanical modeling and sensitivity analysis of bipedal running ability. II. Extinct taxa, J. Morph., (2004) 262: 441-461.

Hutchinson, J.R., Bates, K., T., Molnar, J., Allen, V. and Makovicky, P. J., "Computational analysis of limb and body dimensions in Tyrannosaurus rex with implications for locomotion, Ontogeny, and Growth, PlosOne, (2011) 6: e26037(1-20).

Chapter 13 Mechanical power for 14.1 m/s running

This chapter describes mechanical power calculation and its interpretation of 14.1 m/s running of *T.rex*. This calculation is the first trial in this research area. No one ever accomplished quantitative calculation of mechanical power of running dinosaur not only *T.rex*.

13.1 Mechanical power calculation

Figure 13.1 shows vertical acceleration a_y, and moment of the force of the hip joint M_{hip}, knee joint M_{knee} and ankle joint M_{knee}. The characteristics of 14.1 m/s running motion are described in Sec.11.2. Fig.11.3 displays stick diagram of corresponding running motion. The contribution of the toe joint is omitted because the ankle extensors could have been producing most of the required toe joint moments (Hutchinson, 2004). We observe that the maximum of vertical acceleration is 28.3 m/s², which yields multiplier factor $G=2.89$, where the ground reaction force F is expressed as $F = -m_{body} G g$ with the total mass m_{body} and gravitational constant g.

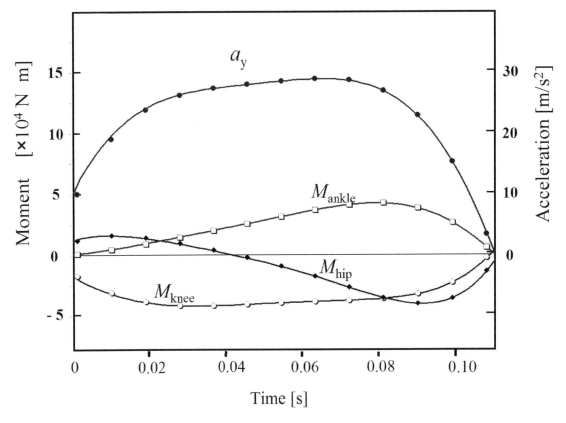

Fig.13.1 The vertical acceleration a_y, and moment of force of each joint M_i where i=hip, knee and ankle. The corresponding running motion in a speed of 14.1 m/s is displayed in Fig.11.3

Chapter 13 Mechanical power for 14.1 m/s running

Absolute value of the maximum absolute value of the moment is obtained as $M_{knee}=4.7\times10^4$ Nm and $M_{hip}=4.7\times10^4$ Nm.

Mechanical power output of each joint P_i is calculated using the relation, $P_i = \omega_i \cdot M_i$. For example, the angular velocity ω_{hip} for a hip joint is the time derivative of the angle θ_{hip}, i.e., $\omega_{hip} = \dfrac{d\theta_{hip}}{dt}$. The definition of angle of each joint is shown in Fig.13.2.

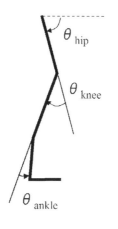

Fig.13.2 The definition of angle for each joint.

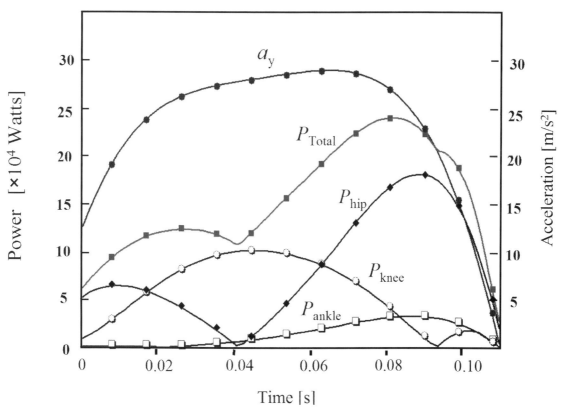

Fig.13.3 Mechanical power for i-th joint P_i, the sum and the vertical acceleration a_y.

Chapter 13   Mechanical power for 14.1 m/s running

Absolute value of each power P_i, total power $P_{total} = P_{hip} + P_{knee} + P_{ankle}$, and vertical acceleration are plotted in Fig.13.3. The mechanical power of the toe joint is omitted as the same reason discussed in the evaluation of the moment of force in Fig.13.1. The maximum power of P_{total} is obtained as 2.4×10^5 (Watts) in the stance phase of 14.1 (m/s) running.

13.2   Mechanical power per kilogram of muscle mass, and comparison with the data of extant animals

For the evaluation of running ability, the mechanical power per kilogram muscle is calculated, and compared to the other data.   The value of P_{total} in Fig. 13.3 is divided by the muscle mass of a leg. As for muscle mass of leg, 16 % of the total mass are employed. The value of 16 % is the maximum ratio derived from recent mass property studies (Bates et al., 2009; Hutchinson et al., 2007). The result is shown in Fig.13.4. For the comparison, the data of extant animal experimentally obtained are shown in the graph. The mechanical power output of muscle of extant animals was described in Chap. 6 in detail.

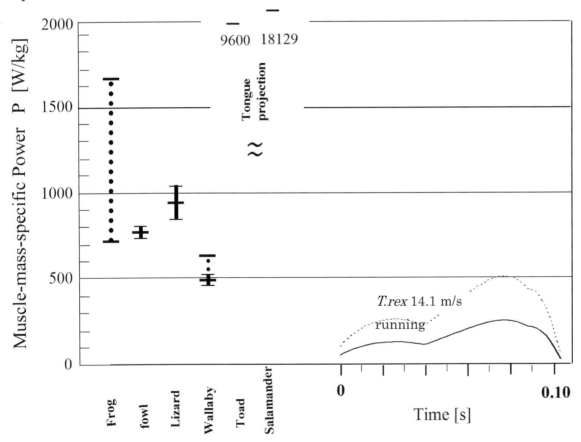

Fig. 13.4   The mechanical power per kilogram of muscle mass (right) and the data for extant animal (left).

Chapter 13　Mechanical power for 14.1 m/s running

Surprisingly, value of the maximum power obtained by experiment is extremely high. Examples of it are the followings;

822-1644 W/kg	frog	(Papelowski and Marsh, 1997)
777.64±32.95 W/kg	fowl	(Henry, et al., 2005)
952±89 W/kg	lizard	(Curtin et al., 2005)
495.0±15.0 W/kg	wallaby	(McGowan et al., 2005)
9,600 W/kg	tongue projection of tod	(Lappin et al. 2006)
18,129 W/kg	tongue projection of salamander	(Deban et al., 2007)

With comparing these data, our simulation value is lower than the maximum value of extant animal. Then, 14.1 m/s running of *T.rex* is thought to be possible on a viewpoint of mechanical power. Otherwise, our simulation result of mechanical power for 14.1 m/s running of *T.rex* does not exceed known maximum of mechanical power of extant animal. It is noted that the bold line shows the power for the case of center of mass being located at hip joint. Twice larger value is shown as dashed line in Fig.13.4. This two fold error estimation is based on miss-location of the hip joint (see Chap.12). Even considering this twofold error estimation, the simulation result of mechanical power is far lower than known maximum power of extant animal.

On the issue of mechanical power of joint that animal generates, a following point is important as is discussed in Chap7.

A mechanism of elastic storage plays a crucially important role to produce mechanical power of the animal. This involves a contribution of the passive element such as tendon. Then, discussion including only muscle is not sufficient to study mechanical power that animal produces.

For the case of the human leg, studies on the role of tendon in movement have been achieved by Fukunaga's group since 2001, (Fukunaga et al., 2001; Kurosawa et al., 2001.; Kurosawa et al., 2003). Tendon is called passive element (Chap.1), which does not generate force. However, muscle is connected to bone through tendon, and this passive element is an important functional unit on producing mechanical power. Such mechanism involving passive element is called elastic storage in the animal experiment. It has been well known that animal achieve extremely high mechanical power with this mechanism. For example, mechanical power of tongue projection of salamander

Chapter 13 Mechanical power for 14.1 m/s running

achieved high mechanical power as 18,129 W/kg, (Daban, et al., 2007). Then, the discussion only involving muscle function is not sufficient on evaluating *T.rex* running ability. Study including passive element is left for us as future work. At present, there is no study on the role of the passive element in producing mechanical power of running motion of *T.rex*.

13.3 Mechanical power calculation for $v=9.8$ m/s running motion

We show the result of the calculation of mechanical power for $v=9.8$ m/s running motion in this section. This running motion is described in Sec.10.3.

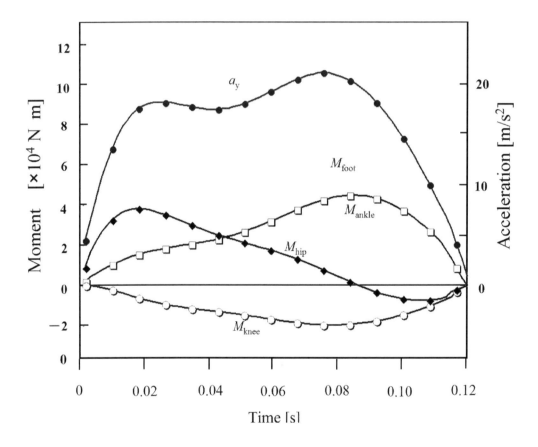

Fig.13.5 The vertical acceleration a_y, and moment of force of each joint M_i where i=hip, knee and ankle. The running motion in a speed of 9.8 m/s is displayed in Fig.10.1

Chapter 13　Mechanical power for 14.1 m/s running

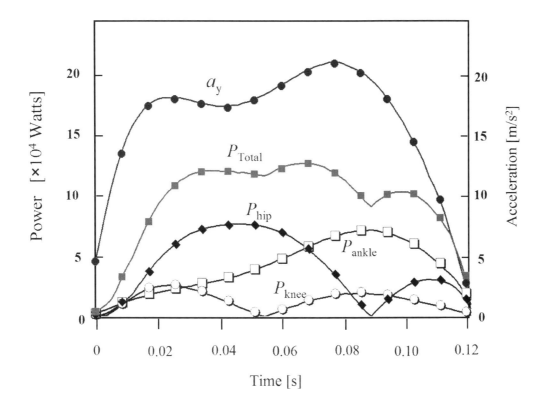

Fig.13.6　Mechanical power for i-th joint P_i, the sum and the vertical acceleration a_y.

For the motion of v=9.8 m/s running, the peak value of the sum of power $P_{total} = P_{hip} + P_{knee} + P_{ankle}$ is 12×10^4 watts, which is half of the case for v=14.1 m/s running (Fig.13.3). Then, running motion in a speed of 9.8 m/s is more plausible for *T.rex* in our simulation. The muscle mass specific power corresponding to the peak of P_{total} is nearly 100 W/kg. It is noted that this range of muscle mass specific power in different taxa have been widely reported as 107 W/kg for mouse (James et al. 1995), 110-122 W/kg for dolphin (Gray, 1936; Weis-Fogh et al., 1977). Then, running motion in a speed of 9.8 m/s is thought to be highly plausible, which is appeared in our running simulation of *T.rex*.

References

Bates, K. T., Manning, P. L., Hodgetts, D. & Sellers, W. I.　Estimating Mass Properties of Dinosaurs Using Laser Imaging and 3D Computer Modelling, *PLoS ONE* 4 (2): e4532 doi:10.1371/journal.pone.0004532(2009).

Curtin, N. A., Woledge, R. C. and Aerts, P., Muscle directly meets the vast power demands in agile lizards, Proc. R. Soc. B (2005 272: 581-584.

Chapter 13   Mechanical power for 14.1 m/s running

Deban, S. M., J.C. O'Reilly, U. Dicke, U. and J.L. van Leeuwen, Extremely high-power tongue projection in plethodontid salamanders. J. Exp. Biol. (2007) 210: 655-667.

Fukunaga, T., Kubo, K., Kawakami, Y., Fukashiro, S., Kanehisa, H. and Maganaris, C.N. 2001. In vivo behaviour of human muscle tendon during walking, Proc. R. Soc. Lond. B (2001) 268: 229-233.

Gray, J. 1936. Studies in animal locomotion. VI. The propulsive powers of the dolphin, Journal of Experimental Biology 13:192-199.

Henry, H. T., Ellerby, D. J. and Marsh, R. L., Performance of guinea fowl *Numida meleagris* during jumping requires storage and release of elastic energy, J. Exp. Biol., (2005) 208: 3293-3302.

Hutchinson, J. R. Ng-Thow-Hing, V. & Anderson, F. C. A 3D interactive method for estimating body segmental parameters in animals: Application to the turning and running performance of Tyrannosaurus rex, *J. Theor. Bio.*   246, 660-680(2007).

Hutchinson, J. R. Biomechanical modeling and sensitivity analysis of bipedal running. *I. Extant Taxa*. J. Morph. 262, 421-440(2004).

James, R. S., Atringham, J. D. and Goldspink, D. F. 1995. The mechanical properties of fast and slow skeletal muscles of the mouse in relation to their locomotory function, Journal of Experimental Biology 198: 491-502.

Kurosawa, S., Fukunaga, T. and Fukashiro, S., Behavior of fascicles and tendinous structures of human gastrocnemius during vertical jumping. Journal of Applied Physiology, (2001) 90: 1349-1358

Kurosawa, S., Fukunaga, T., Nagano, A., and Fukashiro, Interaction between fascicles and tendinous structures during counter movement jumping investigated in vivo, J. Appl. Physiol., (2003) 95: 2306-2314.

Lappin, A. K., Monroy, J. A., Pilarski, J. Q., Zepnewski, E. D., Pierotti, D. J. & Nishikawa, K. C. Storage and recovery of elastic potential energy powers ballistic prey capture in toads. *J. Exp. Biol.*   209, 2535-2553(2006).

McGowan, C. P., Baudinette, R. V. , Usherwood, J. R. and Biewener, A.A., The mechanics of jumping versus steady hopping in yellow-footed rock wallabies, J. Exp. Biol. (2005) 208 :2741-51.

Paplowski, M. M. and Marsh, R. L., "Work and Power Output in the Hindlimb Muscles of Cuban Tree Frogs *Osteopilus Septentrionalis* during jumping, J. Exp. Biol., (1997) 200: 2861-2870.

Weis-Fogh, T., Alexander, R. Mc. N. and Pedley, T. J. 1977. The sustained power output from striated muscle. In, Pedley, T. ed., Scale Effects in Animal Locomotion, p. 511-525. Academic Pres Inc., New York.

Chapter 13 Mechanical power for 14.1 m/s running

Chap 14 Maximal running speed of animals

14.1 Mammalian fast running — Running speed v.s. log(body mass)—

In this chapter, locomotor performance of extant animal is reviewed. Especially, maximal running performance of various taxa of mammals is compared. In the next chapter, maximal running performance of terrestrial reptiles is briefly reviewed.

The reported maximum running speeds of various mammals are plotted semi-logarithmically as a function of body mass in Fig.14.1. The data of Fig. 14.1 are composed of the ones of references (Jones and Lindstedt, 1993; Schenau et al., 1994).

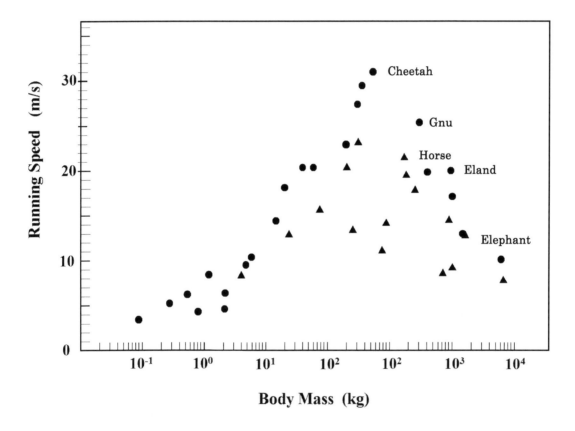

Fig.14.1 Maximum running speeds of mammals plotted as a function of log-body mass. The solid circle represents the data that show large running speed in the work (Jones and Lindstedt, 1993). The solid triangle represents the data that show large running speed in the work (Schenau et al., 1994). Not of all data in these works are shown. For example, data of 148 species of mammals are shown in the work (Jones and Lindstedt, 1993), however, the data showing largest speed for each body mass are only shown.

Chapter 14 Maximal running speed of animals

The data showing maximum running speed for each body mass appeared in the reference (Jones and Lindstedt, 1993; Schenau et al., 1994) are shown in Fig.14.1. As observed from Fig.14.1 maximum running speed does not show a monotonic increase or decrease as a function of body mass. Rather, running speed of cheetah shows the maximum speed over all species of mammals, which is almost middle position in the logarithmic scale of body mass that ranges from 6 g to 6000 kg. The two extremes that are squirrel with 6 g mass and elephant with 6000 kg mass show the smallest value as 8 m/s.

14.2 Running speed with a measure of body length/s

Because muscle shortening length depends on body size, body length-dependent scale is a good quantity to compare running speed and body mass.

Then, we plot semi-logarithmic graph and double-logarithmic graph of speed in body size/s and body mass in Fig. 14.2 and Fig.14.3, respectively.

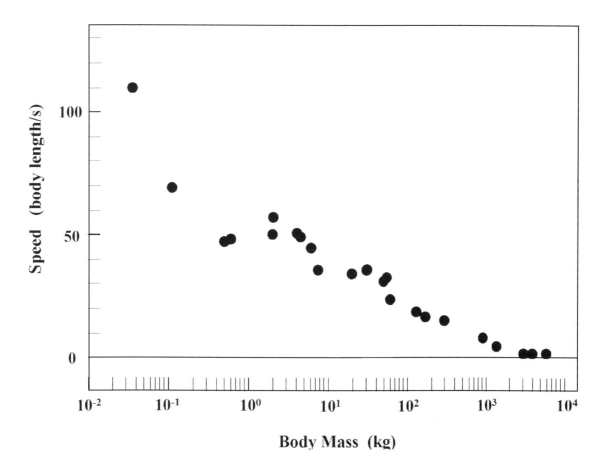

Fig.14.2 Logarithm of body mass v.s. maximum running speed in the unit of body length per second for mammals. The data are the ones of the black circle shown in Fig.14.1.

Chapter 14 Maximal running speed of animals

Figure 14.2 shows the data denoted as black circle in Fig.14.1 in a measure of speed as body length per second. The horizontal axis is the logarithmic scale of body mass. This graph shows the monotonic decrease with the increase of body mass. Interestingly the smallest Merriam kangaroo rat shows the highest speed as 110 body length/s, which is three and a half times faster than cheetah of the speed 32.4 body length /s. The tendency of the data in Fig. 14.2 is monotonic decrease, however, it can not be said that the relation of v in body size/s and the logarithm of body mass is a linear relation. Figure 14.3 shows the same data as double-logarithmic graph. It is also noticed that no single linear relation is observed from this double-logarithmic graph. Then, it means that there is neither of power nor exponential relation between running speed in body size /s and body mass.

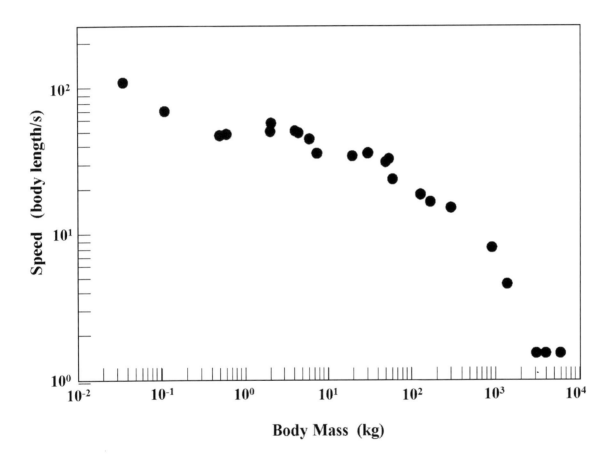

Fig.14.3 Double logarithmic graph of maximum running speed of mammals in the unit of body length/second and body mass. The data are the ones of the black circle shown in Fig.14.1 and Fig. 14.2.

It is noted that Garland searched power low relation as $v_{max} \propto M^{\alpha}$ for mammals of 106 species, where v_{max}, M and α are maximal running speed, body mass and a constant, respectively (Garland, 1983). The result is obtained as $v_{max} \propto M^{0.165 \pm 0.036}$, however, the linear relation in double logarithmic graph does not represent over all data, correctly. Instead, Garland stated that the curvilinear relationship of polynomial regression between $\log_{10} v_{max}$ and $\log_{10} M$ shows a better fit for the data. That is obtained as follows,

$$\log_{10} v_{max} = a + b \log_{10} M - c (\log_{10} M)^2 \quad , \tag{13.1}$$

where a, b and c are constant. The constants are obtained as $a =1.47832$, $b =0.25892$, and $c =0.06237$. This relation relatively well represents overall distribution of v_{max}. However, unfortunately its biological meaning has not been revealed.

On searching analytic relation of running speed and body mass, the following might be said. For each taxonomic group, stride length and frequency of limb is observed, that the product of those gives running speed. Discussion including these quantities may give insight in searching analytic relation of running speed and body mass.

14.3 Fast running performance of reptile

In contrast to the mammalian case, available data of reptile that report fast running is limited. Usually, reptile show slow locomotory performance. Wang et al reported that 0.83 m/s ~ 1.11 m/s were the highest speeds that the lizards were able to maintain on the treadmill (Wang et al., 1997). Carrier reported higher values of the maximum running speed as 2.0 m/s for *Varanus* and 4.5 m/s for *Iguana iguana* (Carrier, 1987). Huey reported 1.59-3.48 m/s running of Kalahari lizard (*Eremias lineoocellata*) whose body weight was 3.41±0.28 g (Huey, 1982). Probably, known maximum speed which has been published in scientific journals is in Garland's report. In the paper, it is described that small lizard *Ctenosaura similes* (iguanid lizard) achieved fast running in a speed of 9.61 m/s (Garland, 1987). The weight of the subject was 230 g. As far as the author concerns, the fastest running speed in the literature is this value.

Although the report was not published on scientific journal, there are several descriptions that lizard achieved fast running comparable to Garland's report. For example, Adam Briton who is the author of crocodilian biology database (http://crocodilian.com) states that most crocodiles can achieve speeds of around 3.3, 3.9 and 4.7 m/s. He denotes as the followings.

"However, crocodiles can accelerate much faster than this over very short distances by exploding into action - I have measured adult saltwater crocodiles (around 4 metres total length) moving at 12 metres per second for a quarter of a second, which is long enough to capture prey standing within one body length before it even has time to react."

Summary

Small iguanid lizard *Ctenosaura similes* achieved fast running in a speed of 9.61 m/s. The body weight of the subject was 230 g.

In an un-authorized document, 12 m/s running for the duration of quarter of a second is noted for 4 metres saltwater crocodiles.

14.4 Running speed estimation of *T.rex* from the data of extant animals.

On the issue of *T.rex*'s fast running ability, we try to evaluate the possibility of fast running of *T.rex* with a relation of extant animal's data that was described in the previous sections. Heavy mammals that achieves fast running in Fig. 14.1 are eland and elephant. An eland with 1000 kg achieves 17 m/s running, and an elephant with 6000 kg achieves 10 m/s running. Interpolation of these data makes possible to the following estimation. An animal with 3000 kg may run in the speed of 12 m/s, which is shown as the black star in Fig. 14.4. It is not be said that an animal with 4000 kg could not run in the speed of 13 m/s, which is shown as the white star in Fig. 14.4. The white square in Fig.14.4 shows the data of an animal of 6000 kg body weight running in a speed of 14 m/s. As observed from extant animal locomotion ability, it may be controversial that this running motion could be realized

Fig 14.4 basically shows running performances of mammals, whereas the rhombuses represent data of reptiles. Iguanid lizard (*Ctenosaura similes*) of 230 g body weight achieved running performance in the speed of 9.61 m/s (Garland, 1984). Kalahari lizard (*Eremias lineoocellata*) of body weight 3.41±0.28 g achieved running performance in the speed range of 1.59-3.48 m/s (Huey, 1982). These data of reptiles show higher running speed compared to mammal for a similar body weight. Then, the possibility is not excluded that *T.Rex* of 6000 kg body weight could run in the speed of 14 m/s.

Finally, it is noted that running performance of *T.rex* of 6000 kg body weight would exceed of the one of human. Known maximal momentary running speed of human is 12.27 m/s which was achieved by Usain Bolt at the 2009 world championships in athletics. The maximum running speed of ordinary person is far slower than this record. It would be in a range of 6 m/s ~ 9 m/s. Then, it is said that *T.rex* of 6000 kg body weight may run faster than human.

Chapter 14 Maximal running speed of animals

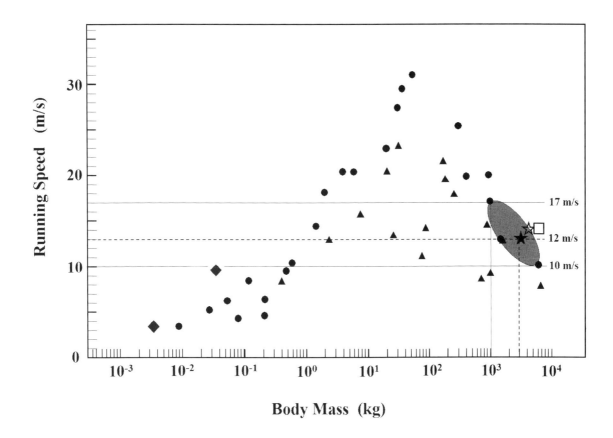

Fig.14.4 The data of maximal running speed of the lizards are shown as rhombuses together with mammals data which are the same of Fig.14.1. Possible running speeds of *T.rex* are shown at lower right in the graph. Those are explained in the text, i.e., black and white star, and white square.

References

Carrier, D. R., Lung ventilation during walking and running in four species of lizards. Exp. Biol. (1987) 47: 33–42.

Garland, T. JR., The relation between maximal running speed and body mass in terrestrial mammals, J. Zool. Lond. (1983) 199: 157-170.

Garland, T. JR., Physiological correlates of locomotory performance in a lizard: an allometric approach. Am. J. Physiol. (1984) 247:R806-15.

Huey, R. B., Phylogenetic and ontogenetic determinants of sprint performance in some diurnal Kalahari lizards. Koedoe (1982) 25: 43-48.

Schenau, G. J. I., Koning, J. J., and Groot, G., Optimisation of sprinting performance in running, cycling and speed skating. Sports Med. (1994) 17(4): 259-275.

Wang, T., Carrier, D. R., and Hicks, J. W., Ventilation and gas exchange in lizards during treadmill exercise. J. Exp. Biol. (1997) 200: 2629-2639.

Chapter 15 Running ability of Tyrannosauridae and scaling property

15.1 Scaling property of theoretical formula

In this chapter, the author discusses on scale dependence of the theory that is appeared in Chap. 9. In Chap. 9 a fraction of muscle mass for i-th joint to the total body mass m_i (%) is introduced as,

$$m_i(\%) = \frac{100 M_i L d}{\sigma c r m_{body} \cos\theta} \quad , \quad [(9.2)], \quad (15.1)$$

which was given by Hutchinson and Garcia (Hutchinson and Garcia, 2003).

Let x be characteristic length of *T.rex*. Mathematically x is equal to a constant α times body length of *T.rex*,

$$x = \alpha \cdot (\text{body length}) \quad , \quad (15.2)$$

This is expressed in physics as,

$$x \sim \text{body length} \quad . \quad (15.3)$$

This means that x is proportional to the body length. Then, each variable appeared in Eq.(15.1) is expressed as follows,

$$\begin{aligned} L &\sim x \\ M_i &\sim x^4 \\ r &\sim x \\ m_{body} &\sim x^3 \end{aligned} \quad (15.4)$$

On the other hand, the following variables are constant as,

$$\begin{aligned} d &= 1.03 \times 10^3 \text{ [kg/m}^3\text{]} \quad &\text{density} \\ c &= 1 \quad &\text{activation ratio} \end{aligned} \quad . \quad (15.5)$$

The other parameters σ and θ are constant, however, its value is controversial or not well investigated,

Chapter 15 Running ability of *Tyrannosauridae* and scaling property

$$\begin{array}{ll} \sigma & \text{the maximum muscle stress} \quad \text{contraversial} \\ \theta & \text{the pennation angle} \quad \text{not well investigated} \end{array} \quad . \tag{15.6}$$

When we insert these into Eq.(15.1), we obtain scale dependence on the fraction of muscle mass as,

$$m_i(\%) \sim x \quad . \tag{15.7}$$

Then, we have a relation between $m_i(\%)$ and m_{body} as,

$$m_i(\%) \sim m_{body}^{1/3} \quad . \tag{15.8}$$

Let us denote the sum of m_i as m, i.e., $m = \sum_i m_i$, which expresses the fraction of a leg muscle mass to the whole mass. For $m = \sum_i m_i$, the same power law of Eq.(15.8) is validated.

$$m(\%) \sim m_{body}^{1/3} \quad . \tag{15.9}$$

Based on the discussion of Chap.11 and Chap.12, let us consider the case that 21 % fraction of muscle mass for a leg to 6000 kg body mass is theoretically required for *T.rex* to run quickly. This fraction is introduced by Hutchinson on the study of *T.rex* running ability (Hutchinson, 2004).

Assume that the $m = \sum_i m_i$ is required as 21% for 6000 kg *T.rex* to run quickly.

Fig.15.1 shows power law graph of $m(\%) \sim m_{body}^{1/3}$. It is noted that Bates et al's mass distribution analysis showed that *T.rex* (MOR555) would have 16 % mass relative to the total mass for one leg (Bates, et al., 2009). Then, only *T.rex* whose mass is smaller than 2600 kg could run, which is shown as the gray area in Fig.15.1. On the other hand, body weight heavier than 2600 kg *T.rex* could not run.

Note that the following values of *m*(%) are published: 14.2 % for MOR 555 (Hutchinson et al., 2007) and 14.4 % for BHI3033 (Bates et al. 2009), and 14.6~25.4 for Carnegie, Sue, Stan and MOR (Hutchinson et al., 2011) . In the next subsection, we will discuss on the meaning of this graph.

Chapter 15 Running ability of *Tyrannosauridae* and scaling property

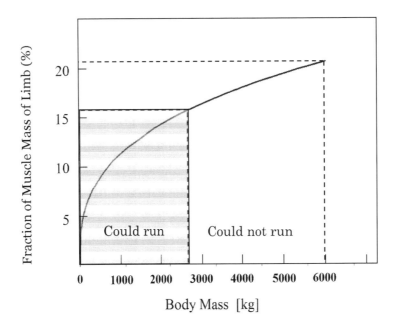

Fig. 15.1 The power law $m(\%) \sim m_{body}^{1/3}$ for a case that required muscle mass fraction $m = \sum_i m_i$ is 21 % for 6000 kg *T.rex*

15.2 Young *T.rex* could run, and adult *T.rex* could not run?

In 2004, Erickson et al published a result that *T. rex* had a large growth rate of 2.1 kg/day. They estimated body weight and age of seven specimen, and drew growth curve of *T. rex* as Fig 15.2.

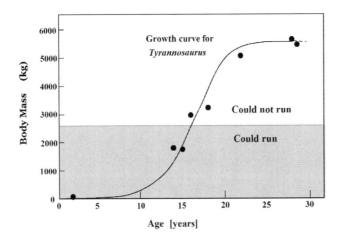

Fig.15.2 Growth curve of *T.rex* based on Erickson et al's data (Erickson et al., 2004). Shadow for lower body mass than 2600 kg is based on Fig.15.1.

If body mass dependence of a fraction of muscle mass of a limb is the one that described in the previous subsection, locomotion pattern of T.rex changes at the middle of the growth process. According to Fig.15.1 and Fig.15.2, transition of locomotion pattern occurred at around sixteen years old. Namely, young T.rex could run, however, adult T.rex could not run. It means that the diet of T.rex changed from carnivore by hunting in young age to carnivore by scavenging in adult age.

Based on this idea, Japanese broadcasting service NHK presented a scenario that a child T.rex in a family chased a prey, and a parent T.rex patched it by ambush attack (Nippon Hoso Kyokai, http://www.nhk-book.co.jp/recommend/science/dinos/dworld_tachiyomi.html). To the author, this scenario sounds unnatural. If such a scenario was true, a question may rise how a single adult T.rex supplied enough food to live. This scenario sounds like that T.rex should form a family to supply enough food. Otherwise, an adult T.rex must die.

Also, the following point can be stated. It is rather natural to consider that both of juvenile and adult T.rex could run. A possible alternative is that adult T.rex could run faster than juvenile T.rex. The reason is that the adult T.rex should need more food than juvenile T.rex. Erickson et al. pointed out that T.rex had a large growth rate of 2.1 kg/day (Erickson et al., 2004), which is clearly observed in Fig.15.2. It is hard to assume that diet of T.rex changed at the middle of this rapid growth period from hunting by running to scavenging by slow walking.

It is noted that the growth rate of 2.1 kg/day corresponds to 767 kg/yr, where yr stands for year (Erickson et al., 2004). Subsequent studies show slower value as 559 kg/yr (Bybee et al., 2006), 601 kg/yr (Erickson et al., 2006). Recent study suggests nearly a half of the first evaluation as 365 kg/yr (Myhrvold, 2013). One report presented far fast growth rate as 1790 kg/yr, however, number of specimens in this study is limited as five (Hutchinson et al., 2011).

15.3 Running ability of a family Tyranosauridae

In this book, only T.rex of 6000 kg body mass has been discussed. In this section, we extend the discussion to bipedal carboniferous dinosaurs of a family Tyrannosauridae. The following genera are known except for Tyrannosaurus in a family Tyrannosauridae, *Albertosaurus, Gorgosaurus, Alioramus, Daspletosaurus, Tarbosaurus, Teratophneus* and *Zhuchengtyrannus*. It is controversial that *Nanotyrannus* is true genus in Tyrannosauridae. Among them, *Albertosaurus, Gorgosaurus* and *Daspletosaurus* are well studied in north America. Erickson et al. drew growth curve of these genus as Fig.15.3. Based on the criterion of running performance of T.rex as in Fig.15.1, all the three genus except for Tyrannosaurus could run, because the weight of *Albertosaurus, Gorgosaurus* and *Daspletosaurus* is less than 2600 kg.

The author thinks that the similarity of bone structure of these genera brought the same locomotion capability. It seems that this is a simple and natural way of thinking. It is difficult to consider that only adult Tyrannosaurus could not run quickly among Tyrannosauridae, and all the others could run quickly.

Chapter 15   Running ability of *Tyrannosauridae* and scaling property

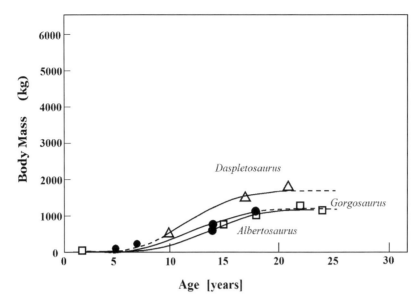

Fig.15.2 Growth curve of three genera of Tyranosauridae, *Daspletosaurus*, *Gorgosaurus* and *Albertosaurus*. This graph is drawn based on Erickson et al's data (Erickson et al., 2004).

Let us consider a case that a required muscle mass fraction of a limb to the body is 16 % for fast running of 6000 kg *T.rex*. The power law relation between body mass and mass fraction becomes as Fig.15.3. In this case, there is no need to consider such a complex and unnatural scenario. Both of juvenile and adult *T.rex* could run based on this assumption, and all genera in Tyranosauridae could run.

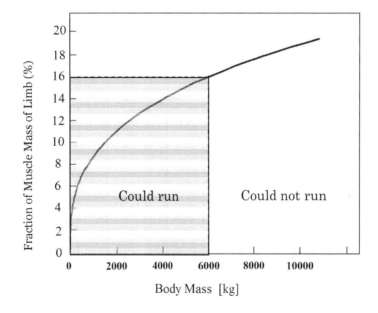

Fig. 15.3   The power law $m(\%) \sim m_{body}^{1/3}$ for a case of $m = 16(\%)$ at 6000 kg body mass.

Chapter 15 Running ability of *Tyrannosauridae* and scaling property

This is very much simple way of thinking. The author believes that this is a natural idea for running ability of *T.rex*.

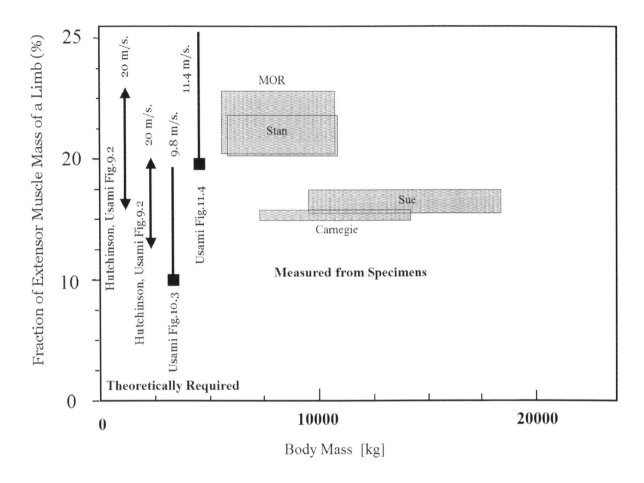

Fig. 15.4 Relation of extensor muscle mass ratio of a limb and body mass. The data denoted at left hand side show theoretically required muscle mass ratio. The data shown as box show range of body mass and muscle mass ratio measured from four specimens (Hutchinson, et al., 2011).

In 2011, Hutchinson et al. built three dimensional computational model of *T.rex* based on mounted skeleton at museum (Hutchinson, et al., 2011). Body mass of some specimen exceeds 6000 kg. They estimated extensor muscle mass ratio of a hindlimb to the whole body mass. The range of possible body mass and extensor muscle mass ratio is shown as box at right hand side in Fig.15.4. Theoretically required extensor muscle mass ratios are summarized at left hand side in Fig.15.4. If measured value of the ratio from specimen is larger than theoretically required value, *T.rex* could run fast. From Fig.15.4, it is observed that the boxes showing measured value locate upper than the minimum value of theoretically calculated one. On the contrary, the maximum values of

Chapter 15 Running ability of *Tyrannosauridae* and scaling property

theoretically required muscle mass ratio exceed the range of boxes. Thus, it is not clearly stated that heavy *T.rex* could or could not run fast from these data. One can not say that *T.rex* could not run fast. Also, one can not say that *T.rex* could run fast based on these data. The data of Fig.10.3 and Fig.11.4 is based on σ =30 N/cm². In biological experiments, the value of σ is reported up to 180 N/cm² as described in Chap.4. When we greater value for σ than 30 N/cm²s, the range of the data based on Fig.10.3 and Fig.11.4 decreases. In this case measured ratios exceed theoretically required value, that means high possibility of fast running of *T.rex*.

Thus, the author of this book has shown possibility that *T.rex* could run in a speed of 9.8 m/s and 14.1 m/s based on known results in biological measurements.

References

Bates, K. T. Manning, P. L., Hodgetts, D., and Sellers, W. I., Estimating Mass Properties of Dinosaurs Using Laser Imaging and 3D Computer Modelling. PLoS ONE (2009) 4 (2): e4532 doi:10.1371/journal.pone.0004532.

Erickson, G. M., Makovicy, P. J., Currie, P. J., Novell, M. A., Yerby, A. A., and Brochu, C. A., Gigantism and comparative life-history parameters of tyrannosaurid dinosaurs, Nature (2004) 430:772-775.

Hutchinson, J. R., and Garcia, M., *Tyrannosaurus* was not a fast runner. Nature (2002) 415: 1018-1021.

Hutchinson, J. R., Biomechanical modeling and sensitivity analysis of bipedal running ability. II. Extinct taxa. J. Morph. (2004) 262: 441-461.

Hutchinson, J. R. Ng-Thow-Hing, V., and Anderson, F. C. A 3D interactive method for estimating body segmental parameters in animals: Application to the turning and running performance of *Tyrannosaurus rex*. J. Theor. Bio. (2007) 246: 660-680.

Hutchinson, J.R., Bates, K., T., Molnar, J., Allen, V., and Makovicky, P. J., "Computational analysis of limb and body dimensions in *Tyrannosaurus rex* with implications for locomotion, Ontogeny, and Growth, PlosOne (2011) 6: e26037(1-20).

Erickson, G. M., Currie, P. J., Inouye, B. D., and Winn, A. A., Tyrannosaur life tables: an example of nonavian dinosaur population biology. Science (2006) 313: 213–217.

Bybee, P. J. and Lee AH, Lamm E-T., Sizing the Jurassic theropod dinosaur Allosaurus: assessing growth strategy and evolution of ontogenetic scaling of limbs. J. Morphol. (2006) 267: 347–359.

Myhrvold, N. P., Revisiting the estimation of dinosaur growth rates, PlosOne (2013) (8.12)e81917(1-24).

Chapter 16 Conclusion

The author states the following conclusions for running ability of *T.rex*.

1. Biological parameters have a very wide range of values. Inserting known maximum value of muscle parameters into the equation of evaluation, it is said that *T.rex* could have mid-stance posture of fast running motion (Chap.4, 9,10,11).

2. Then, we can not exclude the possibility that *T.rex* could not run quickly. In other words, we would say that *T.rex* could run quickly.

3. On the maximum running speed of *T.rex*, there is no theoretical framework to predict it, accurately.

4. Froude number $Fr = \dfrac{v^2}{gL}$ is a useful indicator on discussing running speed of the animal, however, it is a qualitative expression for running speed (Chap.8). It is difficult to predict the maximum running speed of *T.rex* from this expression quantitatively (Chap.11).

5. A method of simulation study unfortunately involves uncertainty of biological parameters and simulation conditions in wide range. Then, no definitive conclusion can be obtained from the simulation study for the maximum running speed of *T.rex*.

6. Results of two computer simulation studies have been published including this work. It tells that *T.rex* could run in speed of 8, 9, ~14 m/s (Sellers et al., 2009: This work) (Chap.10, 11). However, the results are affected by the choice of biological parameters and simulation conditions. Then, it is not definitive conclusion for *T.rex* running ability.

7. Calculation of mechanical power for running motion of *T.rex* is achieved in this work for the first time (Chap.13). The result is in a range of known mechanical power of the extant animal's muscle (Chap.6). Then, it is said that *T.rex* could run quickly based on a discussion of mechanical power.

Chapter 16 Conclusion

8   However, it is said that theoretical study on mechanical power of animal is not fully matured to predict the maximum running speed of *T.rex* precisely. Elastic storage involving tendon plays a crucial role to generate mechanical power of movement of animal (Chap.7). No general theoretical framework involving this effect has been proposed to answer a question such that what is theoretical maximal running speed of *T.rex*. A calculation only involving mechanical power of msucles is insufficient to answer to this question.

9   Specific tension σ is a quantity expressing the force that the muscle can produce at best (Chap.4). It represents the maximum muscle tension for specific muscle groups. Then, the value depends on taxonomic group of animal and muscle groups. Figure 4.9 summarizes the published data of σ which span in a very wide range as 9~180 N/cm².

10 It is noted that extremely high mechanical power 18,129 W/kg was reported by Debian et al. in tongue projection of salamander (Deban et al., 2009).

Known maximum of specific tension is σ =180 N/cm² found in an experiment of the static bite of human mandible. The σ of 153 N/cm² was reported for elbow extensors. The σ =220.0 N/cm² was reported for claw closer muscle of crustacean. Note that, Hutchinson used a value of 30 N/cm² for σ.

Then, experimental studies have revealed high potential of muscle ability.

If we insert known maximum value of muscle property into the evaluation expression for the locomotion of *T.rex*, the result leads to the conclusion that *T.rex* definitely be possible to run faster than human.

Mechanism transferring force using passive element such as tendon enhances muscle ability. However, the function of the total system has not yet been solved satisfactorily. This situation also makes it difficult to predict *T.rex* locomotion ability, correctly.

11 Location of the center of mass in running motion of *T.rex* is an important parameter to calculate a required muscle mass of a leg (Chap.12). If it locates at the hip joint, the required muscle mass becomes smallest (Chap.14), that makes a possibility of fast running of *T.rex* larger.

Chapter 16 Conclusion

12 Many other factors are discussed in detail, for example, moment arm r, leg length L, etc. A choice of these affects the result of running possibility of T.rex. If we assume a certain choice of parameters which is allowed in known biological data, it leads to the high possibility of fast running of T.rex.

If we take a choice of the intermediate value of biomechanical parameters in a range of known values, the result becomes one possible answer among all possibilities that are allowed by published biomechanical data. At present, no one can state definite conclusion to answer the question how fast T.rex could run.

Final conclusion

Predicting the maximum running speed of T.rex accurately based on known biomechanical data is an impossible task, at present, or near future.

Inserting the published maximum value of biomechanical parameters into the evaluation equation leads to the high possibility of fast running of T.rex.

Two simulation studies presented possibility of 8, 9, ~14 m/s running of T.rex, which means that T.rex could run faster than ordinary human. However, it is not definite conclusion.

Then, we can not say that T.rex could not run fast. (T.rex might be a fast runner.)
So, we can not say that T.rex was a scavenger, and always ate dead meat. (T.rex might be a hunter who chased prey by fast running.)

The author of this book considers that T.rex ran fast and hunted prey. This consideration is also supported by the data of running ability of extant animals (Chap.14).

Because of morphological similarity of the bone structure of Tyrannosauridae, the author thinks that all biped dinosaurs of Tyrannosauridae could run quickly. It is difficult to consider that only adult T.rex could not run quickly, and all the others including juvenile T.rex could run quickly (Chap.15).

If only adult T.rex could not run quickly, the way of diet must change at the middle of rapid growth of 2.1 kg/day.

References

Deban, S. M., O'reilly, J. C., Dicke, U., and van Leeuwen, J. L., Extremely high-power tongue projection in plethodontid salamanders, J. Theor. Biol., (2007) 210:655-667.

Sellers, W. I., Manning, P. L., Lyson, T., Stevens, K., and Margetts, L., Virtual Palaeontology: Gait Reconstruction of Extinct Vertebrates Using Hight Performance Computing, Palaeontologia Electronica (2009) 12.3.13A: 1-14.

Appendix Unsolved question on the center of mass of *T.rex*

Simple calculation of the moment of force for a joint

In this appendix, the author presents a simple derivation of the moment of force for i-th joint M_i in Eq. 9.2. Let us consider the simplest model composed of an arm and a sphere. Let us set weight of the arm to be zero, and a sphere whose mass is equal to whole body mass m_{body}. The sphere is small, and attached at the tip of the arm as shown in Fig.A.1.

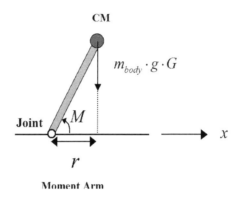

Fig. A.1 A simplest model composed of an arm and a sphere which has whole body mass m_{body}.

For this case the moment of force M for the joint is equal to $r \cdot m_{body} \cdot g \cdot G$, where G is multiplier factor reflecting ground reaction force. In Hutchinson and Garcia's work, this value is set as $G = 2.5$ (Hutchinson and Garcia, 2002). The r and g are moment arm and gravitational constant, respectively. For *T.rex* model, the r is the distance from the joint to the center of mass of *T.rex* in x axis. The value for this variable is set as 0.36m in Hutchinson and Garcia's work (Hutchinson and Garcia, 2002). Then, we have,

$$M = 0.36 \cdot 6000 \cdot 9.8 \cdot 2.5 = 5.3 \times 10^4 \ [\text{N} \cdot \text{m}] \quad , \tag{A.1}$$

as moment of force for the simple model with m_{body}=6000 kg. Best guess value for the moment of force M for each joint of *T.rex*'s limb is shown in Table 14.1. The value of our rough estimation appeared in Eq.(A.1) is comparable to the one of Hutchinson and Garcia's work which is shown in Table A.1.

Appendix   Unsolved question on the center of mass of *T.rex*

	Hip	Knee	Ankle	Toe
M [N · m]	7.5×10^4	-2.4×10^4	6.6×10^4	4.9×10^4

Table A.1   Best guess value for moment of force for each joint of *T.rex* limb in the work (Hutchinson and Garcia, 2002)

The value of the moment of force M appeared in Eq. (A.1) is very close to the one that Hutchinson and Garcia obtained using the expression of Eq.(9.1). It is instructive to consider the case that the center of mass is located just above the foot. For this case, the joint moment M becomes zero,

$$M = 0 \quad [\text{N} \cdot \text{m}] \quad . \quad (\text{A.2})$$

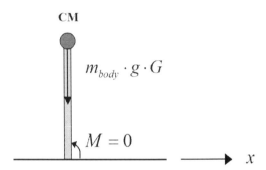

Fig. A.2   When the center of mass is located just above the foot, the joint moment M becomes zero.

Then, the position of the center of mass (CM) is crucially important to estimate joint moment. In chap 9, we assume that the position of CM (center of mass) is located 35 cm cranial from the hip joint. If this distance becomes shorter, a fraction of muscle mass of the limb to the total body mass m becomes far smaller.

Computer aided mass distribution analysis revealed that the center of mass of *T.rex* is located 0.3-0.7m cranial based on mass property study (Bates et al., 2009: Gatesy and Hutchinson, 2007). The increase of this distance makes the large increase of required muscle mass for a leg. Then, a question why the center of mass is not located just above the foot remains as an unsolved question.

Note that Hutchinson et al.'s achieved 3D scanning of five specimens of *T.rex* (Hutchinson et al., 2011), and examined a wide range of whole-body models that included extremes such as too skinny, too fat and too disproportionate ones. They obtained cranial position of center of mass as −0.024~1.160m from hip joint.

Appendix   Unsolved question on the center of mass of *T.rex*

References

Bates, K. T., Manning, P. L., Hodgetts, D. & Sellers, W. I.   Estimating Mass Properties of Dinosaurs Using Laser Imaging and 3D Computer Modelling, *PLoS ONE,* (2009) 4 (2): e4532 doi:10.1371/journal.pone.0004532.

Gatesy, S. M., Baker, M. and Hutchinson, J. R.   Constraint-Based Exclusion of Limb Poses for Reconstructing Theropod Dinosaur Locomotion. J. Vert. Paleo., (2009) 29: 535-544.

Hutchinson, J. R. & Garcia, M. Tyrannosaurus was not a fast runner. Nature, (2002) 415: 1018-1021.

Hutchinson, J. R. Ng-Thow-Hing, V. & Anderson, F. C.   A 3D interactive method for estimating body segmental parameters in animals: Application to the turning and running performance of Tyrannosaurus rex, *J.* Theor. Bio., (2007) 246: 660-680.

Hutchinson, J.R., Bates, K., T., Molnar, J., Allen, V. and Makovicky, P. J., "Computational analysis of limb and body dimensions in Tyrannosaurus rex with implications for locomotion, Ontogeny, and Growth, PlosOne, (2011) 6: e26037(1-20).

Appendix Unsolved question on the center of mass of *T.rex*

Copyright

Cover Page

The photo of black beauty mounted on Royal Tyrrell Museum. Permission is granted from Royal Tyrrell Museum through e-mail march 25, 2014.

Chap. 2.

Fig2.1
Reproduced with permission from Fig.8 of The Journal of Experimental Biology, Milligan et al., 1997, vol.200, pp2425-36.

Fig2.2
Reproduced with permission from Fig.7 of The Journal of Experimental Biology, Milligan et al., 1997, vol.200, pp2425-36.

Fig.2.5
Reproduced with permission from Fig.10 of The Journal of Experimental Biology, Milligan et al., 1997, vol.200, pp2425-36.

Fig.2.6
Reproduced with permission from Fig.8 of The Journal of Experimental Biology, Milligan et al., 1997, vol.200, pp2425-36

Fig.2.7
RightsLink License Number 3216260906674
Jones, D. A., Changes in the force–velocity relationship of fatigued muscle: implications for power production and possible causes, J. Physiol. (2010) 588:2977-286:   Published online 2010 June 14.   doi: 10.1113/jphysiol.2010.190934.

Chap. 4.

Fig. 4.4
RightsLink License Number 3472801508069
Kawakami, Y., et al., Specific tension of elbow flexor and extensor muscles based on magnetic resonance imaging, Eur. J. Appl. Physiol., 1994, vol. 68, pp139-147. (Fig.4)

Copyright

Fig. 4.8
RightsLink License Number 3472801508069
Kawakami, Y., et al., Specific tension of elbow flexor and extensor muscles based on magnetic resonance imaging, Eur. J. Appl. Physiol., 1994, vol. 68, pp139-147. (Fig.5)

Chap. 5.

Fig.5.1
RightsLink License Number 3096260174965
Akagi, R., Takai, Y., Ohta, M., Kanehisa, H., Kawakami, Y., Fukunaga T., Muscle volume compared to cross-sectional area is more appropriate for evaluating muscle strength in young and elderly individuals. Age Ageing. 2009 Sep;38(5):564-9. doi: 10.1093/ageing/afp122. Epub 2009 Jul 13. (Fig.1)

Chap. 6

Fig.6.2
Adapted with permission from Fig.1 of The Journal of Experimental Biology, Josephson, 1985, vol.114, pp493-512.

Fig.6.3
Adapted with permission from Fig.1 of The Journal of Experimental Biology, Josephson, 1985, vol.114, pp493-512.

Fig.6.4
Adapted with permission from Fig.5 of The Journal of Experimental Biology, Askew and Marsh., 2001, vol.204, pp3587-3600.

Fig.6.5
Reproduced with permission from Fig.5 of The Journal of Experimental Biology, Askew and Marsh., 2001, vol.204, pp3587-3600.

Chap. 7.

Fig.7.2
RightsLink License Number 3217121384274
Kurosawa, S., Fukunaga, T. and Fukashiro, S., Behavior of fascicles and tendinous structures of human gastrocnemius during vertical jumping. Journal of Applied Physiology, (2001) 90: 1349-1358.

Printed in Great Britain
by Amazon